The Chondriome – chloroplast and mitochondrial genomes

Edited by S.H. Mantell,
G.P. Chapman and
P.F.S. Street

Copublished in the United States with
John Wiley & Sons, Inc., New York

Longman Scientific & Technical
Longman Group UK Limited
Longman House, Burnt Mill, Harlow
Essex CM20 2JE, England
and Associated Companies throughout the world.

Copublished in the United States with
John Wiley & Sons, Inc., 605 Third Avenue, New York, NY 10158

First published 1986

British Library Cataloguing in Publication Data

The Chondriome: chloroplast and mitochondrial
   genomes.—(Monographs and surveys in the biosciences)
   1. Plant genetics
   I. Mantell, S. H.   II. Chapman, G. P.
   III. Street, P.   IV. Series
   581.1′5   QK981

   ISBN 0-582-46369-6

Library of Congress Cataloging in Publication Data

ISBN 0-470-20699-3 (USA only)

Printed and bound in Great Britain by
Biddles Ltd, Guildford and Kings Lynn

Contents

Preface

Although an extensive morphological diversity is displayed in the phenotypes of eukaryotes, the subcellular organisation of genomes within somatic cells appears to have been conserved throughout the course of evolution. From amphibians to primates and from bryophytes to angiosperms, the majority of DNA within each living interphase cell is localised inside a discrete membrane-bound nucleus. A complex nuclear pore system enables the eukaryotic nucleus to export transcribed genetic information in the form of messenger RNA into the cytoplasm where protein synthesis is located. The cytoplasmic component of eukaryotic cells also contains a retinue of organelles, *i.e.* the mitochondria and the plastids (in the case of plant cells) which contain their own respective types of DNA and their independent protein synthesising systems too. The overall phenotype expressed by the eukaryotic cell is therefore under the integrated control of genetic information based not only in the nucleus but also the cytoplasmic organelles. The genomes of the latter are sometimes collectively referred to as the Chondriome.

It is almost eighty years since the phenomenon of uniparental inheritance was first described in plants. Variegated leaf colour traits in *Pelargonium zonale* were demonstrated to be inherited strictly from the maternal parent whereas flower colour traits present in both sets of parents were inherited in a Mendelian fashion. These results can now be explained: genes encoding for the particular leaf variegation traits were located within the plastids while those encoding for the flower colour traits were located within the nucleus. These types of maternal inheritance in plants appear to be due to the selective exclusion of cytoplasm from the male gametes either prior to, during or after fertilisation. Anomolies are present though. In some genera like *Antirrhinum*, Oenotherea and *Hypericum* a biparental inheritance of plastid traits can be demonstrated. Fascinating questions therefore arise such as why are there apparently different types of chondriome distribution in plants and what mechanisms exist in the male gametophyte to exclude or maintain organelles prior to, during and following fertilisation.

The comparative roles of cytoplasmic and nuclear genes in plants are of particular scientific interest especially now that manipulations of prokaryotic DNA have reached such an advanced stage and that the genetic engineering of crop plants is now gaining momentum with refinements to the Ti- and Ri-plasmid vector systems. Several characteristics of the chondriome require investigation. What is the biological significance of base sequence homology between parts of mitochondrial DNA (mt-DNA) and nuclear DNA in eukaryotes? Why is mitochondrial genome size and structure so plastic compared to those of the chloroplasts which appear to be quite conserved within plants? What are the biological functions of the so-called 'hot-spots' (Active recombinogenic sites within the mitochondrial and chloroplast genomes) and in particular, what role, if any, does the chondriome have for increasing the genetic diversity of crop plants?

Gene mapping of the mitochondrial and chloroplast DNA of a range of important crop species has now been accomplished largely because of the prokaryotic nature of these genomes. Furthermore, the identification and the sequencing of both the promoter and terminator regions of important genes such as those of the ubiquitous large subunit of ribulose bisphosphate carboxylase/oxygenase is proceeding at a rapid rate. The emergence of somatic hybridisation techniques suitable for plants over the last twelve years presents immense opportunities with which to start answering some of the above questions. Protoplast fusion and the regeneration of plants from somatic cell hybrids and gametosomatic hybrids now permits theoretically unlimited combinations of nuclear and cytoplasmic genomes.

It was with these exciting developments in mind that the Second Wye International Symposium chose as its 1985 topic 'The Chondriome'. Fortuitously this event coincided with the quinquagenary of the appearance of a treatise entitled 'Le Chondriome – les constituants morphologiques du cytoplasme' produced by the famous French cytomorphologist Professor A. Guilliermond. The contents of ths current volume therefore log some of the scientific advances made in improving our knowledge of the chondriome over the last fifty years.

S. H. Mantell
G. P. Chapman
P. F. S. Street

Wye College
December, 1985

Acknowledgements

We are indebted to all of the Wye International Symposium contributors who have produced chapters for this volume. Particular thanks are given to Professor Kenton Ko for making a contribution available for publication here.

Finally, we acknowledge the excellent support which was received from all our colleagues at Wye College and from editorial staff at Longman Group Ltd. during the preparation and production of the book.

S. H. Mantell
G. P. Chapman
P. F. S. Street

Publisher's Acknowledgements

We are grateful to the following for permission to reproduce copyright material:

The American Journal of Botany for Figure 4.5 and Tables 4.3 and 4.4; Springer-Verlag (Planta), for Table 4.1

List of contributors

S. Akada
Department of Biological Sciences, University of Maryland, Baltimore County, Catonsville, MD 21227, USA.

T. L. Barsby
Allelix Inc., 6850 Goreway Drive, Mississauga, Ontario, Canada.

L. O. Bjorn
Department of Plant Physiology, University of Lund, Box 7007, S-220-07, Sweden.

C. H. Bornman
Cell and Tissue Culture, Hilleshog Research AB, Box 302, S-261 23, Landskrona, Sweden.

J. F. Bornman
Department of Plant Physiology Univesity of Lund, Box 7007, S-220 07, Lund, Sweden.

G. P. Chapman
Department of Biological Sciences, London University, Wye College, Wye, Ashford, Kent. TN25 6AH, UK.

P. Chetrit
Laboratoire de Photosynthese, CNRS, F-91190 Gif sur Yvette, France.

USDA-ARS/Plant Pathology Department, University of Florida, Gainesville, FL 32611, USA.

E. C. Cocking
Plant Genetic Manipulation Group, Department of Botany, University of Nottingham, Nottingham, NG7 2RD, UK.

S. Cooper-Bland
Department of Biochemistry, Southampton University, Bassett Crescent East, Southampton, SO9 3TU, UK.

H. G. Dickinson
Department of Botany, Plant Science Laboratories, University of Reading, Whiteknights, Reading, RG6 2AS, UK.

A. Hirai
Faculty of Agriculture, Nagoya University, Chikusa, Nagoya 464, Japan.

C. J. Howe
Department of Biochemistry, University of Cambridge, Tennis Court road, Cambridge, CB2 1QW, UK.

N. R. Isola
Plant Pathology Department, University of Florida, Gainesville, FL 32611, USA.

C. Jarl
Cell and Tissue Culture, Hilleshog Research AB, Box 302, S-261 23, Landskrona, Sweden.

R. J. Kemble
Allelix Inc., 6850 Goreway Drive, Mississauga, Ontario, Canada.

K. Ko
Department of Botany, University of Toronto, Toronto, Ontario, M56 1A1, Canada.

X. F. Kong
Department of Biological Sciences, University of Maryland, Baltimore County, Catonsville, MD 21228, USA.

S. D. Kung
Department of Biological Sciences, University of Maryland, Baltimore County, Catonsville, MD 21228, USA.

C. M. Lin
Department of Biological Sciences, University of Maryland, Baltimore County, Catonsville, MD 21228, USA.

R. E. Lloyd
Plant Pathology Department, University of Florida, Gainesville, FL 32611, USA.

P. S. Lovett
Department of Biological Sciences, University of Maryland, Baltimore County, Catonsville, MD 21228, USA.

C. Mathieu
Laboratoire de Photosynthese, CNRS, F-91190 Gif sur Yvette, France.

S. Mongkolsuk
Department of Biological Sciences, University of Maryland, Baltimore County, Catonsville, MD 21228, USA.

G. Pelletier
Laboratoire de Biologie Cellulaire, CNRA, F-78000, Versailles, France.

A. Pirrie
Plant Genetic Manipulation Group, Department of Botany, University of Nottingham, Nottingham, NG7 2RD, UK.

J. B. Power
Plant Genetic Manipulation Group, Department of Botany, University of Nottingham, Nottingham, NG7 2RD, UK.

C. Primard
Laboratoire de Biologie Cellulaire, CNRA, F-78000, Versailles, France.

S. D. Russell
Department of Botany and Microbiology, University of Oklahoma, Norman, OK 73019, USA.

D. Z. Sharpe
Plant Pathology Department, University of Florida, Gainesville, FL 32611, USA.

J. F. Shepard
R. R. No. 1, Terra Cotta, Ontario, Canada.

R. Vankova
Cell and Tissue Culture, Hilleshog Research AB, Box 302, S-261 23, Landskrona, Sweden.

F. Vedel
Laboratoire de Photosynthese, CNRS, F-91190 Gif sur Yvette, France.

J. M. Whatley
Department of Plant Sciences, University of Oxford, South Parks Road, Oxford, OX1 3RA, UK.

S. A. Yarrow
Allelix Inc., 6850 Goreway Drive, Mississauga, Ontario, Canada.

Ultrastructural modifications to organelles during plant development: implications for cytoplasmic gene transfer

1 Patterns of plastid replication during plant development

J. M. Whatley

ABSTRACT. Plastids divide by binary fission at all stages of their
structural development. Proplastids or chloroplasts containing
little or no chlorophyll are responsible for replication during
early organogenesis, at which time the rate of plastid replication
commonly keeps approximate pace with that of cell division. During
later stages of development of the organ, division of the plastids
often outpaces or outlasts cell division and the plastid population
per cell increases. The final size and number of plastids reached
in any cell depends on the type of cell, the organ and the species.
In addition, the overall patterns of plastid replication are clearly
different in the leaves of each of four species investigated.

INTRODUCTION

Before one can satisfactorily tackle the problem of plastid
replication, one must first build up a large amount of information on
the overall pattern of cellular development within the organ with
which one is concerned. The leaves of some species have a persistent
basal meristem; others do not. Successive leaves on the same plant
may follow quite different patterns or timings of development. Any
investigation must take such differences into account. Even within a
single leaf a distinctive pattern of plastid replication is found not
only in each of the six cell layers of the lamina, but also in
different types of cell within a single layer; the mature guard cells,
trichomes, hydathodes and epidermal cells of the primary leaf of
Phaseolus vulgaris all differ from each other in both the size and the
number of plastids per cell (personal observation). Further variation
in plastid numbers is found between different cell types within the
vascular system. As Butterfass (1980) has concluded, the pattern of

3

plastid numbers which is finally achieved depends on several cooperating prepatterns which can be influenced at various speeds and intensities by many genotypic and environmental factors.

Plastids can divide at any state of their structural development from the simplest proplastid to the recently mature chloroplast (Figs. 1a, b, c, d,). Perhaps the greatest problem in studying plastid replication in higher plants is that of obtaining an accurate count of plastid numbers. In some organs, roots for example, and throughout almost all the replication period in other organs, the plastids are small and either lack chlorophyll or contain very little. In consequence, individual plastids are very hard to distinguish under the light microscope. The frequent tendency of these small plastids to cluster tightly round the nucleus further increases the problem.

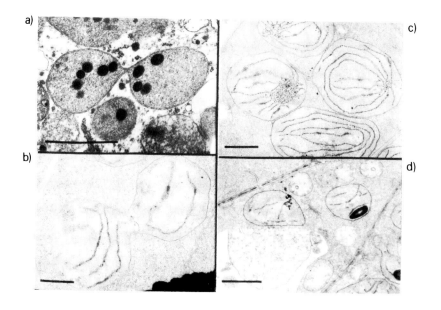

Fig. 1 - a. Proplastid at Stage 1 of structural development
    (*Welwitschia mirabilis*).

  b. Amyloplast; Stage 2 (*Phaseolus vulgaris*).

  c. Pregranal Stage 4 plastid in a Day 5 primary leaf of a light-
    grown plant (*Phaseolus vulgaris*).

  d. Young Stage 5 chloroplast (*Spinacia oleracea*).

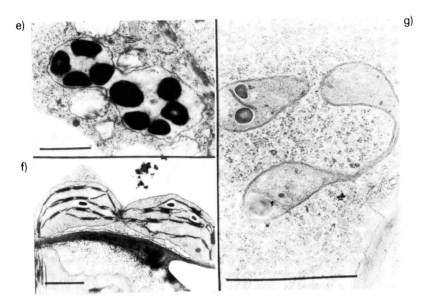

e. Pregranal Stage 4 etioplast in a Day 5 primary leaf of a
   dark-grown plant (*Phaseolus vulgaris*).

f. Two "plastids" which only serial sectioning can show to be part
   of the same dumbell-shaped plastid (*Phaseolus vulgaris*).

g. Two "plastids" which are part of the same Stage 3 pleomorphic
   plastid (*Phaseolus vulgaris*).

The scale bars on all micrographs represent 1 μm.

It is only during those sometimes brief periods when the proplastids or
non-green plastids accumulate starch that their numbers can readily be
determined. Even then, if they are pleomorphic (Fig. 1g), the individual
plastids cannot readily be distinguished: this is a particular problem
during sexual reproduction.

During the early development of an organ, there is considerable mitotic
activity. Any investigation into the changing plastid population must take
account not only of any increase in plastid number per cell, but also of the
increase in number of the cells to which the existing plastid population
must contribute. Almost all investigations on plastid division in higher
plants have avoided this and the other problems by concentrating on the
final stages of the plastid replication period in green or greening leaves
when an accurate count of plastid numbers can be made. At this time,

mitosis is at an end or nearly so, and the developing or recently mature chloroplasts are approaching their maximum size and are lying in the peripheral cytoplasm around an established central vacuole.

There are many and varied patterns of plastid replication to be found in leaves: descriptions of four species of which the author has some knowledge are described below.

ISOETES LACUSTRIS

Leaves of the lower vascular plant genus, *Isoetes,* resemble those of monocotyledons like grasses in that they develop from a persistent basal meristem. In *Isoetes* spp., each meristematic cell contains a single large plastid and close coordination of the processes of cell division and plastid division ensures that a daughter plastid is transmitted to each daughter cell. In leaves of *Isoetes lacustris* (Whatley, 1974a), the mother plastid becomes pleomorphic and then closely encircles the nucleus and divides shortly before or during the early part of prophase (Fig.2a).

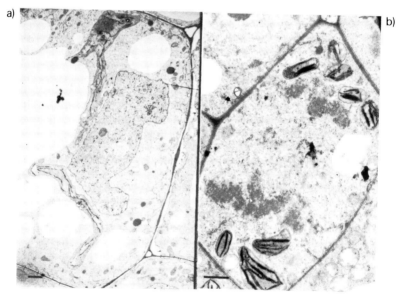

Fig. 2 - a. A pre-prophase cell in the basal meristem of a leaf of
       Isoetes lacustris showing the two daughter plastids close to
       the nucleus, but already moving towards the spindle poles.
    b. Two recently separated daughter cells; each has a plastid
       still at a former spindle pole.

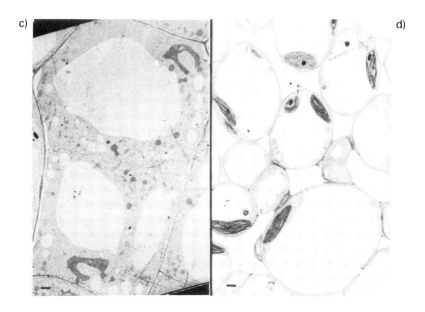

c. A dividing cell in a young leaf (15 mm) of *Theobroma cacao*.
At the upper spindle pole are at least 5 plastids; at the
lower pole are at least 3 plastids.

d. An older leaf (90 mm) of *Theobroma cacao*; each cell contains
3 chloroplasts.

By metaphase the daughter plastids have already migrated to opposite
poles of the dividing cell, where they remain until cytokinesis is completed
(Fig. 2b). In leaves of other species of *Isoetes*, the relative timings of
nuclear and plastid division appear to be similar (Marquette, 1907; Ma,
1928), though in meristems elsewhere in the *Isoetes* plant separation of
the daughter plastids may be delayed until metaphase (Stewart, 1948).

Each leaf of *Isoetes* arises from a single apical cell. Many cycles of
cell division must take place before the final complement of leaf cells is
reached. (Twenty doubling cycles would be needed to produce just 1 million
cells). During each of these many cycles the closely controlled relationship between cell and plastid division is maintained. Furthermore during
each cell cycle the mother plastid and, later, the daughter plastids
consistently take up their distinctive positions, first, encircling the
nucleus and, later, at the two spindle poles.

When cell division has stopped, the plastids continue to divide.

Ultrastructural modifications to organelles

Immediately above the meristem, cells with 1, 2 and 4 plastids can be found. Towards the tip of the leaf (the most mature part), cells with as many as 30 plastids have been observed (Ma, 1928; Stewart, 1948). Stewart has made the interesting observation that in the first cycle of plastid division after cell division has stopped, the dividing plastid continues to encircle the nucleus and, although the nucleus does not divide, the daughter plastids later move to opposite ends of the cell. Subsequently, in the last 4 cycles of plastid division (from 2 to about 32 plastids per cell), the positioning of the plastids seems at all times to be random and the last coordinating link between plastid and nucleus is lost. Such phasing out of the dual control system over the final few cycles of plastid division seems also to be characteristic of the division process in other species.

Butterfass (1980) has pointed out that the nature of the coordination between cell division and plastid division remains uncertain. One type of division may stimulate the other or, perhaps more likely, a third factor might promote both types of division. The factor(s) which promote the cessation of division are equally unknown. Butterfass suggests that the dosage of nuclear DNA or some part of it may have an important direct influence on plastid division and he points out that higher ploidy is associated with higher plastid numbers. However, as he also emphasises, this alone cannot account for the variations in plastid populations which exist. It would nevertheless be interesting to know if, in the *Isoetes* leaf, the extra cycles of plastid division which take place after cell division have ceased to occur in conjunction with the cells becoming endopolyploid.

PHASEOLUS VULGARIS

Both the cotyledons and the primary leaves of *Phaseolus vulgaris* are initiated during the period of seed development on the parent plant. In the primary leaf, all six cell layers of the lamina are established well before seed ripening begins. During this preliminary phase, development is far from synchronous and the precise duration of the growth period between the establishment of the flattened lamina and the onset of seed ripening is variable. For convenience, this pre-ripening growth period has been indicated in the Figures as lasting 10 days. The second phase of development of the primary leaf follows germination and here the precise timing of events is much easier to establish. Meristematic activity continues for about a week after the start of germination and, during the same period, several sets of trifoliate leaves and, soon afterwards, the floral primordia are initiated.

For the primary leaf, Dale (1964) found that cell division stopped first in the upper epidermis (by day 5), but continued in the palisade cells, where it ceased entirely by day 9, at which time the leaf had attained 17% of its final size. Unlike *Isoetes*, there was no persistent basal meristem. Indeed there was no significant difference between the tip and the base of the leaf at the time when cell division stopped. In my own work with a different cultivar of *Phaseolus*, I have removed the seed coat prior to planting. This has not only promoted synchrony of development, but has also resulted in earlier emergence of the seedling and an earlier spurt of lamina enlargement. The different timings for the cessation of cell division and the numbers of cells formed in these seedlings were, however, similar to those reported by Dale, but the leaves had reached 34% of their final size when mitotic activity ended about day 8.

As a leaf develops, a number of distinctive but overlapping events take place. Though it is not possible to separate development into discrete periods of a) cell division, b) differentiation and c) expansion, it is clear that the cessation of cell division is followed in orderly sequence by the establishment of the maximum thickness of the lamina and, later, of maximum surface area. Though some minor differences in size and structure between the cells of the six layers of the lamina can already be identified prior to seed ripening, the major period of cellular differentiation follows germination (Whatley, 1979). How, then, do the patterns of plastid replication fit within the overall mosaic of patterns of leaf development? Counts of the numbers of chloroplasts per cell in the mature leaf show considerable variation between different types of cell (*c*.20 in upper epidermal, *c*.45 in palisade and *c*.32 in mesophyll cells). At what stage of leaf development can differences in plastid populations first be identified?

As pointed out in the Introduction, a direct count of numbers using the light microscope is seldom possible during the proplastid stages of development. In the primary leaf of *Phaseolus vulgaris*, the plastids remain as proplastids throughout the entire period of growth prior to seed ripening and for the first few days of germination. Chloroplasts with grana appear

for the first time about day 6 (Whatley, 1974b). Though proplastids are easy
to identify in thin sections of tissue using the electron microscope, counts
of their numbers per thin section do not, except under special circumstances,
provide an accurate guide to their total numbers per cell (Fig. 1f, 1g). The
method of assessment which I have used (Whatley, 1980) has been to multiply
the mean number of plastid sections per transverse cell section (Fig. 3a) by
the mean number of plastids intercepted by a line through the cytoplasm
parallel to the long axis of cells cut in longitudinal section (Fig. 3b).
Though this method is fairly crude, the plastid counts obtained are in
reasonable agreement with those few direct counts which were possible - as
it was on day 2 of germination during a peak of starch accumulation, and
following day 6 when the chloroplasts mature.

Differences between plastid populations in cells of the upper epidermis
and upper palisade layers of the primary leaf of *Phaseolus* can first be
identified prior to seed ripening (Fig. 4). The greater increase in number
of plastids per palisade cell at this time could be the result either of a

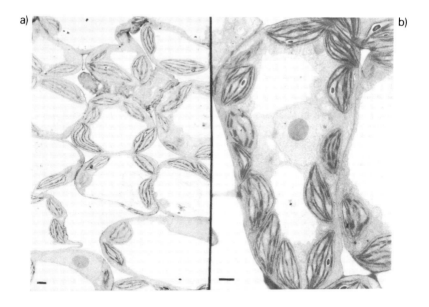

Fig. 3 - a. Palisade cells cut in transverse section (*Phaseolus vulgaris*
             primary leaf, Day 8)

        b. Palisade cells cut in longitudinal section (*Phaseolus vulgaris*
             Day 8).

c)

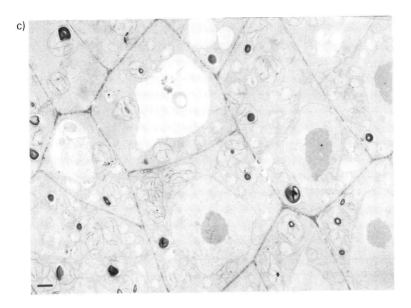

c. A thin section of randomly cut palisade cells with the arrowed
   plastids showing various stages of division (*Phaseolus vulgaris*
   primary leaf, Day 5).
The scale bars on all micrographs represent 1 μm.

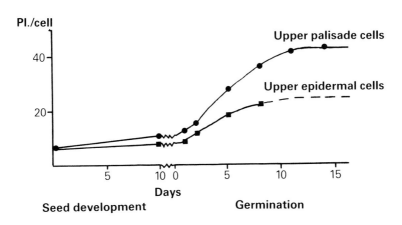

Fig. 4 - The number of plastids per upper epidermal cell and per
        upper palisade cell in the primary leaf of *Phaseolus*
        vulgaris during seed development and following germination.

more rapid rate or prolonged period of plastid division prior to the onset of seed ripening. At the start of germination the plastid population in both types of cell increases rapidly but after only a few days the difference between the plastid populations in the two cell layers becomes much greater. This is principally because plastid division (like cell division) persists for longer in the palisade cells.

A description of plastid replication based solely on changes in average populations per cell can, however, be highly misleading. It provides no information about overall rates of plastid division and the late conspicuous increase in the plastid population in palisade cells indicated in Figure 1 tends to obscure the importance of the earlier phase of plastid replication when plastid division essentially keeps pace with cell division. To follow the overall pattern of plastid replication, one must consider the process in terms of cycles of replication of both cells and plastids, and in terms of total numbers of plastids in the whole of each particular cell layer, a task for which much of the basic information is lacking. Nevertheless one can, even today, provide a preliminary description of the changes which take place.

Lack of synchrony has hindered detailed investigation of the first period of primary leaf development prior to seed ripening. For this reason, a simple flattened lamina of 3000 cells (500 cells in each of the six layers) was select as a convenient starting point. The end point of this phase of growth is the seed prior to ripening, in which the primary leaf contains about 1×10^6 cells, still equally distributed (c. 165,000 cells) in each of the six layers. Following germination the number of cells in the leaf increases to 45-50 $\times 10^6$ (Dale, 1976). Different timings and rates of cell division result in the establishment during this period of development of different numbers of cells in each layer; in the mature leaf the upper palisade layer contains about 15 $\times 10^6$ cells and the upper epidermis 4.5-5 $\times 10^6$ cells. During this more synchronous period of growth, the changes in cell and plastid number can be followed in more detail (Whatley, 1980). From the information obtained I have calculated the number of cell doubling cycles which take place within the upper epidermal and upper palisade layers. By following the changes in plastid number per cell, the total number of plastids in each layer can also be calculated and the mean daily rates of cell and plastid cycling can be estimate These calculations depend, however, on the assumption that all cells and all plastids divide.

There is evidence that in young leaves of several species, the increase in cell number is approximately exponential (Dale, 1976). Dale considers that in the *Phaseolus* primary leaf the proportion of cells undergoing division is close to 100% during the early stages of germination though the proportion later declines. The assumption that all plastids divide will be discussed later. Though the calculations represent an obvious oversimplification of events, they can, nevertheless, provide a useful guide to the patterns of replication which are followed in the developing leaf.

It has thus been estimated that the cells in each layer of the primary leaf undergo just over 8 sequential cell doubling cycles during seed development on the parent plant. In the same period, plastids in the upper palisade layer undergo about 9 doubling cycles. Following germination the palisade cells perform between 6 and 7 doubling cycles and their plastids between 8 and 9 cycles (Fig. 5). The equivalent numbers for the upper epidermal layer during germination are about 5 cell and between 5 and 6 plastid cycles. In all, therefore, upper palisade cells double in number between 14 and 15 times and their plastids about 17 times. (For cocklebur, *Xanthium pennsylvanicum*, the figures used by Maksymowych (1973) indicate a total of 16 cell cycles for the leaf as a whole during the equivalent period of development).

Thus in the palisade cells of the primary leaf of *Phaseolus vulgaris*, as in the leaf of *Isoetes lacustris*, the plastids undergo several cycles of division more than do the cells. However, in *Isoetes* the 5 extra cycles of plastid division are confined to the period after cell division has stopped, whereas in *Phaseolus* the 3 or so extra cycles are spread over a period from before to after cell division comes to an end at day 8 (Fig. 5).

A logarithmic plot of the changing number of plastids within the entire upper palisade layer of the primary leaf of *Phaseolus* shows that plastid division is rapid throughout seed development and for several days after the start of germination (Fig. 5). The mean rate of plastid doubling reaches a peak of about 2 cycles per day between days 2 and 5. When plastid division has been considered only in terms of the increase in plastid number per cell (Figs. 4 and 5), it has sometimes led to the erroneous conclusions a) that the fastest rates of division must be found during the final phase of replication when the increase in number per cell is at its peak, and b) that the earlier stages represent a lag phase (eg. Honda, Hongladarom-Honda,

Ultrastructural modifications to organelles

Kuanyuen and Wildman 1971). It is often forgotten that - if all plastids divide - 50% of the entire complement of plastids within, say, the mature leaf, will be formed during only one cycle of doubling, the final cycle.

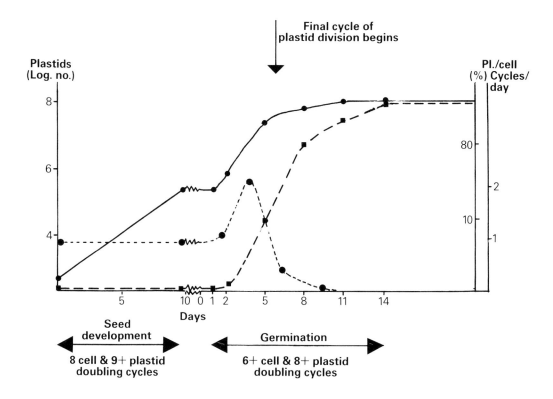

Fig. 5 - A logarithmic plot (●——●) and the percentage (■– –■) of the total number of plastids and the mean daily rate of plastid doubling (●– –●) in the upper palisade layer of the primary leaf of *Phaseolus vulgaris* during seed development and following germination.

Thus, the later few but more obvious cycles of division by larger plastids which can more readily be followed in the light microscope tend to obscure the importance of the much more numerous but less easily identified doubling cycles carried out by smaller proplastids or young chloroplasts. It should be noted, too, that in *Phaseolus*, the final but numerically preponderant 2 or 3 cycles of plastid division only begin when the rate of plastid doubling has already begun to decline (Fig. 5). It may well be that in other species also, the final few cycles of replication which are most often investigated represent the tailing off of the division process and may, as was suggested for *Isoetes*, therefore be atypical.

A further question relating to plastid division is whether or not the same patterns of plastid replication are repeated in successive leaves. Many developmental processes show a pattern of acropetal drift - the appearance of particular developmental features at progressively later stages in successive leaves or other organs. In the photosynthetically active roots of the water fern, *Azolla pinnata*, for example, root hair formation, xylem differentiation and several other cellular and chloroplast features all show this phenomenon (Gunning, 1978; Whatley and Gunning, 1981). In *Phaseolus vulgaris*, too, some features of chloroplasts in successive leaves also appear to show acropetal drift. The time interval between leaf initiation and the cessation of plastid division in palisade cells seems to increase in the successive trifoliate leaves (Fig. 3). A population of 15 and of 35 plastids per palisade cell is attained at a later stage of leaf expansion in the third trifoliate leaf than in the primary leaf, though a final population of more than 40 plastids per palisade cell is attained by both. However, the proportion of the replication period in which division is carried out by proplastids is very high in the primary leaf (16 out of the 17 cycles), but becomes progressively lower in each successive leaf. Though information about plastid numbers in the trifoliate leaves is very limited, it is nevertheless sufficient to suggest that, as one might expect, the patterns of plastid replication in successive leaves cannot be assumed to be identical.

SPINACIA OLERACEA

A considerable amount of work has been carried out by Possingham and his associates on plastid replication in cells of the 4 mesophyll layers in spinach leaves. Possingham's group has not only made itself familiar with the basic process of leaf development, but has taken care to distinguish between results obtained from successive leaves (Chaly and Possingham, 1981; Possingham, 1976; 1980; Possingham and Saurer, 1969; Saurer and Possingham, 1970; Scott and Possingham, 1980; 1983).

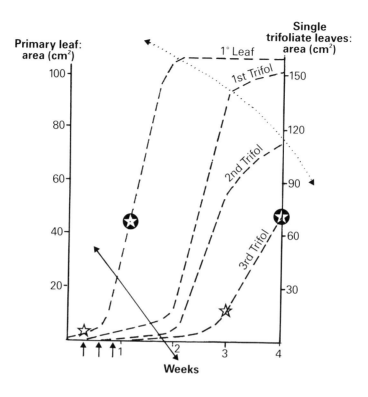

Fig. 6 – The development of successive leaves of *Phaseolus vulgaris*
showing:

a. the time of initiation of 3 sets of trifoliate leaves (↑)

b. the time of transformation of pre-granal to granal
chloroplasts (◄━►).

c. the stage of development of the primary and third trifoliate leaves when the mean number of plastids per palisade
cell reaches 15 (☆) and 35 (★).

d. the approximate time of cessation of plastid division in each
leaf (◄┅►).

The leaves of *Spinacia*, like those of *Isoetes*, have a persistent basal
meristem (Saurer and Possingham, 1969). As in *Phaseolus*, but not as in
Isoetes, the plastid population per cell increases both before and after
cell division stops (Scott and Possingham 1980; 1983). In *Spinacia* one
finds not just a gradient of progressively larger numbers of plastids per

cell between the base and the tip of the leaf, but a gradient which itself is continuously changing. For this reason, it is much more difficult to work out the overall pattern of plastid replication in the spinach leaf than it is in either *Isoetes* or *Phaseolus*.

If one takes, as with *Phaseolus*, an initial lamina of 3000 cells (2000 cells in the four main photosynthetic mesophyll layers) as a convenient starting point, then the final number of 15.8×10^6 mesophyll cells (Possingham, 1980) represents about 13 cell doubling cycles. However, one must bear in mind that in the spinach leaf with its basal meristem, the calculated final number of doubling cycles represents an average figure which takes no account of the changing rates of cycling as cells progress towards the leaf tip. The increase in plastid number from about 10 in young meristematic cells (Chaly and Possingham, 1981) to about 200 per cell in the 100mm leaf (Possingham, 1980), requires between 4 and 5 additional plastid doubling cycles.

Even in very small leaves (less than 2mm in length and containing about 40×10^3 mesophyll cells) the dividing plastids, though small, already contain grana (Fig. 1d). Thus in leaves of spinach, unlike the primary leaf of *Phaseolus*, most division (12 or more of the 17 to 18 doubling cycles) is carried out by plastids which are already chloroplasts.

The Possingham group have also investigated the changes in amount of plastid DNA in cells in different parts of developing leaves. They have found that the number of genome copies per plastid increases from 125 to 200 between the base and the middle of a 20mm leaf (the early phase of plastid replication), and that the amount of plastid DNA per cell then reaches a peak level at which it remains although plastid division still continues. Thus, during the late phase of plastid replication the number of genome copies per plastid falls - to 110 copies at the tip of the 20mm leaf and to 32 copies at the tip of older 100mm leaves (Scott and Possingham, 1983). A similar increase followed by a decrease in genome copies of DNA per plastid has been estimated for cells of increasing age along the length of wheat leaves, though in wheat a peak of over 1000 genome copies per plastid is reached (Boffey and Leech, 1982). These observations are, perhaps, another indication that the final few cycles of plastid division represent a tailing-off of the replication process and are atypical of the main phase of division.

Ultrastructural modifications to organelles

THEOBROMA CACAO

The leaves of *Theobroma cacao*, like those of several other tropical species, do not become conspicuously green until leaf expansion is complete or nearly so. The late greening could reflect either a delay in chlorophyll synthesis and chloroplast maturation, or a delay in chloroplast growth and division. This problem was investigated by Baker and Hardwick (1973), who found that active chlorophyll synthesis was delayed until the later stages of leaf expansion. There was no increase in chloroplast number during this time, and each mesophyll cell in the mature leaf contained only 3 chloroplasts (Baker, Hardwick and Jones, 1975).

A population of 3 plastids per cell is also characteristic of younger leaves (Fig. 2d) - those which have reached less than a quarter of their final length (personal observation). However, meristematic cells in very small leaves (up to 15mm in length) seem to have a larger plastid population, one which is still undergoing division (Fig. 2c). In this very young leaf tissue examined under the elctron microscope, up to 9 discoid plastid sections per cell section have been counted. Though the counts were based solely on this sections, it was clear that dividing cells often contained more than the 6 plastids which might be expected if a cycle of plastid division were immediately to precede one of cell division, and so give rise to daughter cells each containing 3 plastids (Figs. 2c and 2d).

In *Theobroma cacao*, the plastid population per mesophyll cell thus apparently declines during the very early stages of leaf development, though there must be a substantial overall increase in the number of plastids within the leaf as a whole. Subsequently, when cell division comes to an end, the plastid population remains constant at 3 per cell. Although so little is known about the plastids in *Theobroma* leaves, it is clear that the pattern of plastid replication in relation to leaf growth is completely different from that in the species described above (Fig. 7).

PETALS

Most multicellular marine algae tend to have their chloroplasts concentrated in the outermost layers of cells. By contrast, in land plants which receive light unfiltered by water, the plastids in the superficial, epidermal cells

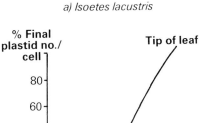

a) *Isoetes lacustris*

b) *Phaseolus vulgaris*

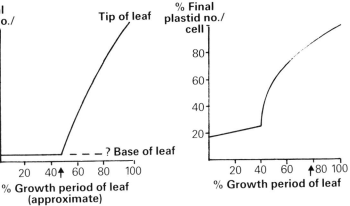

c) *Spinacia oleracea*

d) *Theobroma cacao*

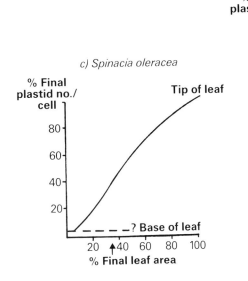

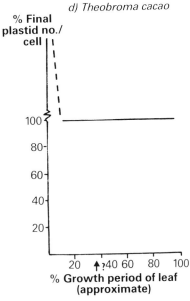

Fig. 7 – Generalised graphs suggesting the varied patterns of increase in plastid numbers per cell during leaf development in four species. Arrows indicate the end of cell division and broken lines represent projected trends.

    a. *Isoetes lacustris*. Division of the single plastid initially keeps pace with cell division; later, plastid division continues after cell division has stopped.

    b. *Phaseolus vulgaris*. Plastid division initially keeps pace with cell division but later outpaces and outlasts it.

     c. *Spinacia oleracea.* Plastid division outpaces and outlasts cell division in the middle and upper parts of the leaf; plastid division in the basal meristem approximately keeps pace with cell division for some time but the pattern of plastid replication which follows the cessation of cell division in these basal cells is not known(?).

     d. *Theobroma cacao.* Plastid division initially outpaces cell division.

of most organs exposed to light are usually few and small whereas the chloroplasts in the cells below are many and larger (Fig. 8a and 8b). This programming of plastid size and number related to cell type seems to be reversed in the flower petals of those species in which the carotenoid content is high. In *Caltha palustris*, for example, the plastids (chromoplasts) in the upper epidermis are numerous and large (Plate 8c) whereas the chromoplasts of the mesophyll cells are few and very small (Whatley, 1984).

    The carotenoid-rich chromoplasts in petals of the yellow form of a pansy *(Viola tricolor)* cultivar have a pattern of distribution similar to

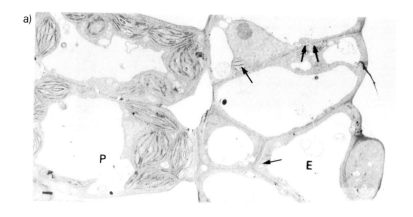

Fig. 8 - a. Cells of the upper epidermal (E) and upper palisade (P) layers of the primary leaf (*Phaseolus vulgaris*, Day 7). Plastids in the epidermal cells (arrowed) are few and small. They occupy only a very small proportion of the cell volume.

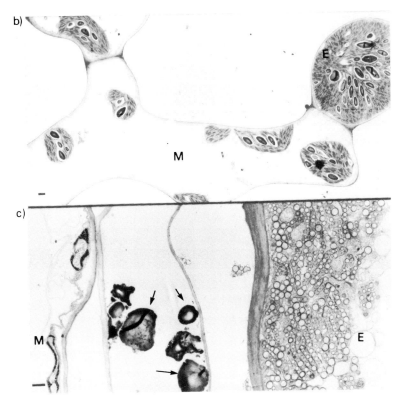

b. In leaves of the lower vascular plant, *Selaginella kraussiana*,
   each upper epidermal cell (E) contains a single, large, cup-
   shaped chloroplast (cut here transversely through its base)
   which, unusually among land plants, occupies a large proportion of the cell volume. The mesophyll cells (M) below
   each contain several smaller platids.

c. The upper epidermal cells (E) in the petals of *Caltha*
   palustris contain many large pleomorphic chromoplasts; the
   lower mesophyll cells (M) have only a few small chromoplasts.
   Starch grains (arrows) are all that remain of the plastids in
   a specialized light-reflecting layer of cells below the
   epidermis.

The scale bars on all micrographs represent 1 μm.

that in *Caltha*. The purple form of the same *Viola* cultivar has
anthocyanin (located in the vacuoles) as the principal pigment, but a small

amount of the same carotenoid (as in the yellow form) is present in the plastids. However, in the purple petals, there are very few chromoplasts in the epidermal cells (personal observation). The pre-pattern for plastid replication must therefore be very different in the two forms of this cultivar. However, no accurate information about plastid repli-cation in petals is available for any species. The tendency of petal plastids to be pleomorphic will continue to make it hard to obtain reliable estimates of their numbers.

Chromoplasts can develop directly from simple proplastids and from amyloplasts or they can develop from chloroplasts following dedifferentia-tion to the proplastid state and subsequent redifferentiation. Almost nothing is known about the DNA in chromoplasts. However, Herrmann (1982) reported that the buoyant density of DNA in chromoplasts of *Narcissus pseudonarcissus* was the same as that in leaf chloroplasts. For the same species, Liedvogel (1976) reported that the number of genome copies per leaf chloroplast was 53 and per petal chromoplast was only 9.

OTHER MODIFICATIONS OF PLASTID NUMBER

The patterns of plastid replication so far considered have, with one exception, involved examples in which the rate of plastid division keeps pace with or exceeds that of cell division or in which plastid division continues after cell division has stopped. Are there any other examples known in which there is a reduction in plastid number per cell and if so how is this reduction accomplished?

It is now well established that during the early mitotic divisions of sporogenesis in mosses, the number of plastids in each cell is reduced from several in the archesporial cells to one in the pre-meiotic sporocyte (eg. Mueller, 1974; Brown and Lemmon, 1980a, b). It is not however clear if plastid division comes to a complete halt while cell division proceeds or if plastid division continues but at a much slower rate than cell division, or even if, as suggested by Jensen and Hulbary (1978), the reduction in number is accomplished by plastid fusion. No substantiated example of a similar decline in plastid population in a vascular plant being specifically associated with different rates or timings of cell and plastid division is known to the author. Yet such procedures might be expected and may, indeed, operate.

In the root of the water fern, *Azolla pinnata,* the number of plastids in the single apical cell apparently declines with age (Whatley and Gunning, 1981). The rate of plastid division may gradually become slower as the apical cell becomes older and so become insufficient to keep up with the rate at which plastids are incorporated into succesive daughter cells. On the other hand, the leaf buds and leaves of *Theobroma cacao* described above may provide an example of the cessation of plastid division prior to that of cell division (Figs 2c and 2d). However, both these examples are based on superficial observations and on counts made from thin sections examined with the electron microscope and so require much more thorough investigation. Another aspect of such differential rates and times of cell and plastid division may take place in some angiosperm species in which plastids are inherited from both parents. In the embryo, the two sets of inherited plastids may divide at different rates or may begin division at different times and so influence the final pattern of distribution of the plastids within the developing plant (Hagemann, 1976).

Changes in plastid number per cell are not, however, necessarily the result of changes in the rates and timings of cell and plastid replication. Other plant responses too can alter, at least temporarily, the number of plastids in a cell. Though these responses can take place at many stages of the life cycle, they seem to be most conspicuous during sexual reproduction.

It is now established that plastids from the (+) and (-) gametes in the green alga, *Chlamydomonas* and some of its close relatives, fuse following fertilization (reviewed in Whatley, 1982). It has also been suggested that a reduction in plastid number by fusion may be achieved by some land plants, *inter alia* by the anthocerote *Megaceros flagellaris* during regeneration (Burr, 1969) and in some mosses during sporogenesis (Jensen and Hulbary, 1978). However, no example of plastid fusion seems to have been substantiated either for proplastids or for chloroplasts in healthy tissues of any land plant.

A major change in plastid number can be achieved by the division of a cell in which the plastids are asymmetrically distributed. In the megaspore mother cells of the gymnosperm *Ginkgo biloba* (Stewart and

Ultrastructural modifications to organelles

Gifford, 1967) and of the angiosperm *Zea mays* Russell, 1979) for example, the plastids become concentrated towards the base. They remain in this position when the mother cell divides to form the linear tetrad of spores. All, or nearly all, the plastids are thus concentrated in the basal - functional - megaspore. In the uninucleate pollen grain of *Phaseolus vulgaris* Whatley, 1982), the plastids become concentrated in the part of the cell farthest from the site of the future generation cell. As a result, following asymmetrical cell division, the plastids tend to be excluded from the generative cell and its successors, the male gametes. The concentration of plastids at a precise position within a particular type of cell or at a particular time during a cell's life is a relatively common event (*cf. Isoetes lacustris*), but the means by which it is carried out is not clear. In moss sporocytes prior to meiosis, microtubules seem to participate in the precise tetrapolar alignment of the plastids (Brown and Lemmon, 1980a, b; 1982), but no such close association with microtubules has been observed in higher plants.

Another means of modifying the plastid population is by an asymmetrical division of a cell in which the plastids are evenly distributed in the cytoplasm. In this way each of the two daughter cells receives a different number of plastids. During regeneration of new plants from mature gametophytes of the anthocerote, *Megaceros flagellaris*, the first few divisions are carried out by cells initially containing about 4 plastids. These divisions are asymmetrical and such that one of the daughter cells always receives only 1 plastid. This single plastid condition is then maintained during several subsequent cycles of cell division (Burr, 1969).

Few plant cells completely lack plastids; gametes, particularly male gametes, are the most common of those which do. Elimination of the plastids can take place either before, sometimes long before, or after fertilization. Several different elimination mechanisms may operate sequentially, none necessarily being completely successful on its own. One of the most successful elimination mechanisms used to promote maternal inheritance of the plastids is found among lower land plants. In many bryophytes and lower vascular plants, the paternal plastid (or plastids) is first moved to the rear of the developing gamete and, later, is removed when a cytoplasmic vesicle containing the plastid is discarded by the sperm before it reaches the egg. Plastids can also be eliminated as a result of their degeneration or digestion. In most taxonomic groups it is the paternal plastids which may be eliminated in this way; exceptionally,in some conifers, it is the maternal plastids in the egg cell which are so removed (Whatley, 1982).

Although the various mechanisms described above do indeed seem to play an important role in modifying and controlling plastid populations, we have only the most sketchy information about them. A great deal of dedicated research will be necessary before we can even begin to understand how these systems operate.

THE PROPORTION OF PLASTIDS WHICH DIVIDE

The series 1, 2, 4, 8, 16 and about 32 chloroplasts per cell observed in submerged leaflets of the moss, *Sphagnum cuspidatum* (Butterfass, 1980) and of 1, 2, 4 etc. plastids observed in leaves of the lower vascular plant, *Isoetes* spp. indicate that all the plastids in each cell can undergo division. Is this true also for higher plants?

When a plastid divides, it commonly undergoes a preliminary phase of elongation. This is followed by the formation of a constriction, usually median, which marks the future plane of division. The constriction becomes progressively tighter until, just before separation, the mother plastid assumes a dumbell shape, the two plastid halves being linked only by a narrow isthmus, which may itself become encircled by an electron dense annulus (Chaly and Possingham, 1981; Leech, Thomson and Platt-Aloia, 1981; Suzuki and Ueda, 1975; Whatley *et al.*, 1982). For larger chloroplasts, the various phases of division can be seen with the light microscope, but, for smaller plastids, the electron microscope is needed (Figs. 1a, 1b, 1c, 1d). Even then, the most deeply constricted dumbell and separation phases of division can only be distinguished from each other with difficulty at the lower magnifications which one must use if the data are to be treated statistically.

Estimates of the proportion of plastids undergoing replication which are obtained only from single thin sections will always be much too low; the necessary distinguishing criteria (constriction or isthmus) are frequently not included in the plane of section and two well-separated, apparently non-dividing plastid figures may well be two halves of one dumbell (Figs. 1f and 3c). Serial thin sections provide a more reliable estimate of the percentage of plastids which were dividing at the moment of sampling, but the work is time-consuming and the numbers of plastids included within any one series tend to be low.

Recent preliminary work on leaves of *Phaseolus vulgaris* and *Spinacia oleracea* suggests that though the plastids within each cell do not divide in synchrony, the proportion of plastids undergoing replication can vary greatly over a 24h period and that there are times when most plastids in most cells are at one or other of the division phases (Whatley and Possingham, unpublished observations).

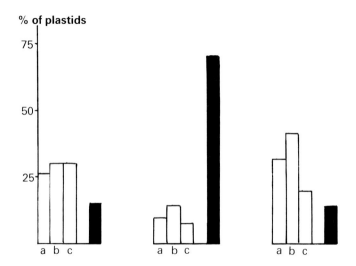

Fig. 9 - The proportions of plastids which are not dividing (black)
or are at different phases of the division process (white).
a. elongation
b. constriction
c. dumbells and recently separated pairs of daughter plastids.

Serial sections of cells at the base of 2.5-3.0 mm first and second leaves of *Spinacia*, grown under a regime of 14-h light and 10-h darkness, showed that at 9.30 am, 30 minutes after the beginning of the light period, 86% of the plastids were dividing (Fig. 9). Of the 31 cells examined only 2 lacked any dividing plastids. By contrast, at 2.00 pm, still during the light period, only 29% of the spinach plastids were at one or other of the designated phases of division. (A high proportion - 87% - of dividing plastids was also observed in five day old primary leaves of *Phaseolus vulgaris* shortly after the beginning of the light period).

Thus, although plastid division was not precisely synchronous, there was nevertheless at least one peak period during the day when most plastids were replicating. These observations would seem to be in agreement with light microscope observations, i.e. when division is actively in progress, all plastids in most cells enter the division process. However, an accurate count of any change in plastid number would be needed to provide evidence that division was actually completed.

THE EFFECTS OF LIGHT ON PLASTID DIVISION

Plastid replication can be influenced by many factors, both internal and external, but one of the most important is light. It is commonly stated that though chloroplast division requires light, proplastids can undergo several division cycles in darkness (eg. Butterfass, 1980).

In the apices of roots of *Phaseolus vulgaris*, non-green plastids within the first 12 tiers of the root proper apparently undergo about 2 cycles of replication in excess of the cycling required to keep pace with cell division; those within the first 12 tiers of the root cap undergo 1 extra cycle (Whatley, 1983). Thus, even in the dark, the rate of proplastid division can surpass that of cell division (Whatley, 1983). In isolated roots of *Convolvulus arvensis*, it was found that at a distance of 200mm from the tip, plastid sections per cortical cell section were 5 times more frequent in light-grown than in dark-grown roots (Heltne and Bonnett, 1970). This suggests that in roots, as in other plant organs, plastid division can be stimulated by light.

Following germination of seeds from parent plants grown under a normal regime of alternating light and darkness, cell number in the primary leaf of *Phaseolus vulgaris* increases from 1×10^6 to $12-18 \times 10^6$ in darkness and to $45-50 \times 10^6$ in the light (Dale, 1976). In the dark, plastid number per palisade cell increases to a maximum of about 25 (Bradbeer, Gyldesholm, Smith, Rest and Edge, 1947b), the same number as is attained by palisade cells after 5 days of growth under a regime of 12h light and 12h darkness (Whatley, 1980). It is interesting that during this same five day period, the plastids in both light and dark-grown

plants follow the same course of structural development and have reached
a superficially similar pre-granal stage. Although only the light-
grown plants contain chlorophyll, both sets of plastids contain single
perforated lamellae and bithylakoids or incipient grana (Plates 1c, e).
The etioplasts in dark-grown plants also contain small prolamellar bodies
which increase considerably in size if growth in the dark is continued.
If the young dark-grown plants are transferred to the light, the
etioplasts are quickly transformed into chloroplasts and plastid numbers
increase. However, if the period of growth in darkness is prolonged,
then, on transfer to light, greening takes place very slowly and the
plastids do not undergo further division (Bradbeer, Gyldenholm, Ireland,
Smith, Rest and Edge, 1974a).

Assuming that the proportion of upper palisade cells within the
primary leaf of *Phaseolus vulgaris* is the same in both light- and dark-
grown plants, then, during germination, the overall increase in plastid
number within this cell layer in darkness is between 5 and 6 doubling
cycles, only about 2 cycles fewer than are carried out in the light.
In spinach leaf discs cultured with sucrose as an energy source,
chloroplasts increased in number from 20 to 60 per cell during 9 days
of growth in darkness (Possingham, 1980). It may be relevant that,
during germination, the movement of sugars and other metabolites from
the cotyledons to the primary leaves of *Phaseolus* begins to decline at
about the same time as plastid division in dark-grown plants comes to
an end.

Replication of mature grana-containing chloroplasts of both vascular
and non-vascular plants normally requires light and the rate of division
is influenced both by daylength and by light intensity up to a
saturating level (Asahi and Toyama, 1982; Possingham, 1976). Although
there have been several investigations of the effects of light of
different wavelengths on plastid division, the responses remain far from
clear.

It has been suggested that the process of plastid division includes
a light-sensitive and a light-insensitive phase (Kass and Paolillo, 1974).
Under low-intensity (0.2-0.6 mW cm^{-2}) white or green light, spinach
plastids increase in size and become deeply constricted, but do not
complete the separation process (Possingham, 1976). On the other hand

there is evidence to indicate that the actual separation of daughter plastids can take place in darkness, though they fail to begin the elongation and constriction phase of a new cycle (Kass and Paolillo, 1974), Possingham, 1976). This suggests that there may indeed be two light-sensitive phases, the first requiring light of low intensity, promoting growth and constriction, and the second requiring light of higher intensity, perhaps needed to bring the connecting isthmus to a state of readiness for separation of the daughter plastids.

PLASTID NUMBERS AND LEAF AREA

The factors responsible for the cessation of plastid division are not known though the process has been shown to cease abruptly soon after cell expansion has stopped (Boasson, Bonner and Laetsch, 1971). The space within a cell which is occupied by plastids is determined by both number and the size of the plastids. It has been proposed that, for each species, the plastids in each type of cell continue to multiply and to increase in size until they occupy a characteristic proportion of the cell volume or surface area (Honda et al., 1971).

The relationship between the cessation of plastid division and the attainment of maximum cell volume has on occasion been misinterpreted as also applying to the attainment of maximum leaf area. Certainly in the primary leaf of Phaseolus vulgaris, as in the leaves of many other species, plastid replication and leaf expansion both stop at about the same time, but the later stages of leaf enlargement are marked by extension of the air spaces between the cells, including the palisade cells, rather than by increasing the girth of the cells.

In the Phaseolus leaf, the palisade cells increase in width at the start of germination, but temporarily decrease in width about day 5 (Fig. 3c; Fig. 6) as the cells begin to increase rapidly in height and the lamina becomes thicker. Further increase in cell width comes to an end by day 7 or day 8 (Fig. 8a), by which time the leaf has reached only 34% of its final area. Then the central vacuole is well-established and the chloroplasts are lined up within the peripheral cytoplasm. From this time onwards, the number of chloroplasts seen in transverse cell sections remains consistently between 4 and 5 (Fig. 3a), but the number of plastids intercepted in longitudinal cell sections (Figs. 3b and 4a)

Ultrastructural modifications to organelles

continues to increase slowly until day 14, at which time longitudinal
extension of the cells also comes to an end (Fig. 10) and the cells
have attained their maximum volume. The increase in palisade cell
volume and cell surface area during the final stages of plastid replication is therefore related to increasing cell height, i.e. the result
of growth in a direction at right angles to the leaf surface; the fact
that leaf enlargement and plastid division in palisade cells both end at
the same time thus appears to be coincidental.

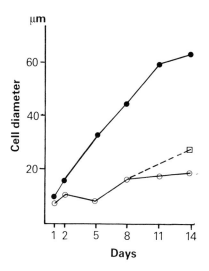

Fig. 10 - Changes in the diameter of upper cells in the primary leaf
        of *Phaseolus vulgaris* during germination.
        cell height (●——●)
        true cell width (O——O)
        apparent cell width if no allowance is made for the air
        spaces between cells (□-- --□).

CONCLUSIONS

During the early stages of development of a plant organ, both cells and
plastids perform many cycles of division. The onset of such a series of
cycles of plastid replication is closely coordinated with the onset of cell
replication. This appears to be true not only in those cells of lower plants
which initially contain only one plastid, but also in higher plant cells

containing several plastids (Fig. 11). As might be expected, there is in monoplastidic lower plant cells greater synchrony of cell and plastid division and the distribution of plastids to daughter cells is more strictly controlled (Figs. 2a, 2b and 2c).

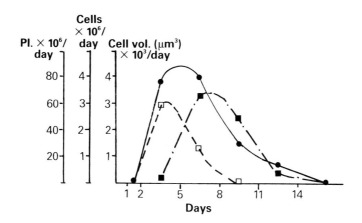

Fig. 11 - The mean daily rate of increase in cell number (□-- --□)
plastid number (●——●) and cell volume (■--·-■) in the upper
palisade layer of the primary leaf of *Phaseolus vulgaris*
following germination.

In the leaves of some land plants, plastid replication tends at first to proceed at approximately the same rate as cell replication but, as meristematic activity comes near its end, coordination between the two replication systems seems gradually to be lost. Plastid division outpaces cell division and/or it continues for several extra cycles. At the same time, replication of the plastid DNA may stop, with the result that the number of genome copies per plastid declines. However, a different regime of cell and plastid division seems to be followed in the leaves of *Theobroma cacao* (Fig. 7).

Throughout the entire replication period, additional control systems appear to operate independently of each other within each separate cycle of plastid division. They influence not only the replication of plastid DNA, but also each successive phase of the plastid division process - elongation, constriction, separation etc. Plastids can divide at any stage of their development up to and including the young grana-containing chloroplasts. However some of the controls affecting replication during the early proplastid stages of development differ from those affecting the later chloroplast stages. Light, for example, appears to be required for one or two steps during the division process in chloroplasts, but is not essential for proplastid division. The nature of these control systems remains obscure. Sucrose can apparently replace the light requirement for chloroplast division. Kinetin has been shown to stimulate further division in mature leaf chloroplasts which are no longer replicating, when discs from the leaves are cultured in an appropriate medium (Boasson, Bonner and Laetsch, 1971).

The period of plastid cycling seems to come to an end at about the same time as cells reach their maximum volume (Figs. 5 and 11). Within each species, both the sizes and the numbers of plastids per cell can be very different in different types of cell (Figs. 8a, 8b and 8c; Fig. 4). In the *Phaseolus vulgaris* primary leaf, palisade cells reach their maximum volume and plastid division stops when the seedlings are about 14 days old, but the plastids continue to increase in size for several more days (Whatley, 1980). In the upper epidermis both plastid division and plastid growth stop appreciably earlier and the plastids remain small and fewer in number. Nevertheless, it is possible that, under fixed growing conditions, the plastids come to occupy a characteristic proportion of the final cell volume or surface area, criteria which in turn may, as Butterfass (1980) suggests, reflect both the dosage of nuclear DNA and the programmed pre-patterning of cell development. However the information so far available about patterns of plastid replication merely serves to illustrate our general ignorance of this whole subject.

REFERENCES

Ashi, Y. and Toyama, S. (1982). Some factors affecting chloroplast replication in the moss *Plagiomnium trichomanes*. *Protoplasma 112*, 9 - 16.
Baker, N.R. and Hardwick, K. (1973). Biochemical and physiological aspects of leaf development in cocoa (*Theobroma cacao*). I. Development of chlorophyll and photosynthetic activity. *New Phytol. 72*, 1315 - 1324.

Baker, N.R., Hardwick, K. and Jones, P. (1975). Biochemical and physiolo-
    gical aspects of leaf development in cocoa (*Theobroma cacao*). II
    Development of chloroplast ultrastructure and carotenoids. *New
    Phytol*. *75*, 513 - 518.

Boasson, R., Bonner, J.J. and Laetsch, E.M. (1971). Induction and
    regulation of chloroplast replication in mature tobacco leaf tissue.
    Plant Physiology 49, 97 - 101.

Bradbeer, J.W., Gyldeholm, A.O., Ireland, H.M.M., Smith, J.W., Rest, J.
    and Edge, H.J.W. (1974a). Plastid development in primary leaves of
    Phaseolus vulgaris. VIII. The effects **of** the transfer of dark-grown
    plants to continuous illumination. *New Phytol*. *73*, 271 - 279.

Bradbeer, J.W., Gyldeholm, A.O., Smith, J.W., Rest, J. and Edge, H.J.W.
    (1974b). Plastid development in the primary leaves of *Phaseolus
    vulgaris*. IX. The effect of short-light treatments on plastid
    development. *New Phytol*. *73*, 281 - 290.

Boffey, S.A. and Leech, R.M. (1982). Chloroplast DNA levels and the
    control of chloroplast division in light-grown wheat leaves.
    Plant Physiology 69, 1387 - 1391.

Brown, R.C. and Lemmon, B.E. (1980a). Ultrastructure of sporogenesis in a
    moss, *Ditrichum pallidum*. I. Meiotic prophase. *The Bryologist 83*,
    137 - 162.

Brown, R.C. and Lemmon, B.E. (1980b). Ultrastructure of sporogenesis
    in a moss, *Ditrichum pallidum*. II. Metaphase I through the tetrad.
    The Bryologist 83, 153 - 160.

Brown, R.C. and Lemmon, B.E. (1982). Ultrastructure of sporogenesis in the
    moss, *The Bryologist 83*, 153 - 160.

Brown, R.C. and Lemmon, B.E. (1982). Ultrastructure of sporogenesis
    in the moss, *Amblystegium riparium*. I. Meiosis and cytokinesis. *Amer.
    J. Bot. 69*, 1096 - 1107.

Burr, F.A. (1969). Reduction in chloroplast number during gametophyte
    regeneration in *Megaceros flagellaris*. *The Bryologist 72*, 200 - 209.

Butterfass, Th. (1980). The continuity of plastids and the differentiation
    of plastid populations. *In Chloroplasts* (ed. Reinert, J. *Results and
    problems in cell differentiation Vol. 10* pp. 29-44) Springer Verlag
    Berlin, Heidelberg, New York

Ultrastructural modifications to organelles

Chaly, N. and Possingham, J.V. (1981). Structure of constricted proplastids in meristematic plant tissues. *Biology of the Cell 41*. 203 - 210

Dale, J.E. (1964). Leaf growth in *Phaseolus vulgaris*. I. Growth of the first pair of leaves under constant conditions. *Ann. Bot. 28,* 579 - 589.

Dale, J.E. (1976). Cell division in leaves. *In Cell division in higher plants* (ed. Yeoman, M.M. pp. 315-345). Academic Press, London, New York, San Francisco.

Gunning, B.E.S. (1978). Age-related control of the number of plasmodesmata in cell walls of developing *Azolla* roots. *Planta 143,* 181 - 190

Hageman, R. (1976). Plastid distribution and plastid competition in higher plants and the induction of plastome mutations by nitrosourea-compounds. *In Genetics and biogenesis of chloroplasts and mitochondria* (eds. Bucher, T., Neupert, W., Sebald, W. and Werner, S.). pp. 331 - 388. North Holland, Amsterdam.

Heltne, J. and Bonnett, H.T. (1970). Chloroplast development in isolated roots of *Convolvulus arvensis* (L). *Planta 92,* 1 - 12.

Herrmann, R.G. (1972). Do chromoplasts contain DNA? II. The isolation and characterization of DNA from chromoplasts, chloroplasts, mitochondria and nuclei of *Narcissus*. *Protoplasma 74,* 7 - 17.

Honda, S.I., Hongladarom-Honda, T., Kwanyuen, P. and Wildman, S.G. (1971). Interpretations of chloroplast reproduction derived from co-relations between cells and chloroplasts. *Planta 97,* 1 - 15.

Jensen, K.E. and Hulbary, R.L. (1978). Chloroplast development during sporogenesis in six species of mosses. *Amer. J. Bot. 65,* 823 - 833.

Kass, L.B. and Paolillo, D.J. Jr. (1974). The effect of darkness and inhibitors of protein synthesis on the replication of chloroplasts in the moss, *Polytrichum*. *Zeitschrift fur Pflanzenphysiologie 73,* 198 - 207.

Leech, R.M., Thomson, W.W. and Platt-Aloia, K.A. (1981). Observations on the mechanisms of chloroplast division in higher plants. *New Phytol. 67,* 1 - 9.

Liedvogel, B. (1976). DNA content and ploidy of chromoplasts. *Naturwissenschaften 63,* 248.

Ma, R.M. (1928). The chloroplasts of *Isoetes melanopoda*. *Amer. J. Bot. 15,* 277 - 284.

Maksymowych, R. (1973). Analysis of leaf development. In *Developmental and cell biology* (eds. Abercrombie, M., Newth, D.R. and Torrey, J.G.). Vol. 1 pp. 1 - 109. Cambridge University Press.

Marquette, W. (1907). Manifestations of polarity in plant cells which apparently are without centrosomes. *Beihefte zum botanschen Zentralblatt 21*, 281 - 303.

Mueller, D.M.J. (1974). Spore wall formation and chloroplast development during sporogenesis in the moss *Fissidens limbatus*. *Amer. J. Bot. 61*, 525 - 534.

Possingham, J.V. (1976). Controls to chloroplast division in higher plants. *J. de Microscopie et de Biol. Cellulaire 25*, 283 - 288.

Possingham, J.V. (1980). Plastid replication and development in the life cycle of higher plants. *Ann. Rev. Plant Physiol. 31*, 113 - 129.

Possingham, J.V. and Saurer, W. (1969). Changes in chloroplast number per cell during leaf development in spinach. *Planta 86*, 186 - 194.

Russell, S.D. (1979). Fine structure of megametophyte development in *Zea mays*. *Can. J. Bot. 57*, 1093 - 1110.

Saurer, W. and Possingham, J.V. (1970). Studies on the growth of spinach leaves (*Spinacia oleracea*). *J. Ex. Bot. 21*, 151 - 158.

Scott, N.S. and Possingham, J.V. (1980). Chloroplast DNA in expanding spinach leaves. *J. Ex. Bot. 31*, 1081 - 1092.

Scott, N.S. and Possingham, J.V. (1983). Changes in chloroplast DNA levels during growth of spinach leaves. *J. Exp. Bot. 34*, 1756 - 1767.

Stewart, K.D. and Gifford, E.M. Jr. (1967). Ultrastructure of the developing megaspore mother cell of *Ginkgo biloba*. *Amer. J. Bot. 54*, 375 - 383.

Stewart, W.N. (1946). A study of the plastids in the cells of the mature sporophyte of *Isoetes*. *Bot. Gaz. 110*, 281 - 300.

Suzuki, K. and Veda, R. (1975). Electron microscope observations on plastid division in root meristematic cells of *Pisum sativum* (L) *Jap. Bot. Mag. 88*, 319 - 321.

Whatley, J.M. (1974a). The behaviour of chloroplasts during cell division of *Isoetes lacustris* (L) *New Phytol. 73*, 139 - 142.

Whatley, J.M. (1974b). Chloroplast development in primary leaves of *Phaseolus vulgaris*. *New Phytol. 73*, 1097 - 1110.

Whatley, J.M. (1979). Plastid development in the primary leaf of *Phaseolus vulgaris:* variations between different types of cells. *New Phytol. 82,* 1 - 10.

Whatley, J.M. (1980). Plastid growth and division in *Phaseolus vulgaris, New Phytol. 86,* 1 - 16.

Whatley, J.M. (1982). Ultrastructure of plastid inheritance: green algae to angiosperms. *Biol. Revs. 57,* 527 - 569.

Whatley, J.M. (1983). The ultrastructure of plastids in roots. *Int. Rev. Cytol. 85,* 175 - 220.

Whatley, J.M. (1984). The ultrastructure of plastids in the petals of *Caltha palustris* (L) *New Phytol. 97,* 227 - 231.

Whatley, J.M. and Gunning, B.E.S. (1981). Chloroplast development in *Azolla* roots. *New Phytol. 89,* 129 - 138.

Whatley, J.M., Hawes, C.R., Horne, J.C. and Kerr, J.D.A. (1982). The establishment of the plastid thylakoid system. *New Phytol. 90,* 619 - 629.

2 Organelle selection during flowering plant gametogenesis

H. G. Dickinson

ABSTRACT   The possibility is explored that changes taking place during microsporogenesis operate to select a proportion of the cell organelles for transmission to the next generation. With the aid of data from numerous developmental stages in several plants it is possible to show that microsporogenesis is accompanied by a phase of organelle dedifferentiation and elimination followed, in the surviving organelles, by a rapid synthesis of DNA. Evidence is presented that a different form of selection mechanism may also be operating in the Compositae where, in addition to elimination of organelles, a spectacular interaction takes place between the major part of the mitochondrial population and the nuclear envelope.   Details of this active, and other more passive systems, are considered in terms of plant breeding strategies and in particular the occurrence of cytoplasmic male sterility in plant populations.

INTRODUCTION

Whilst the transmission of the male and female meiotic products to the next generation is central to the sexual reproduction of flowering plants, zygotic fusion also involves the transmission of organelle populations. These organelles are commonly regarded as being derived from the egg cell rather than the sperms but for many years evidence has been available, the best known being from *Pelargonium* (Hagemann, 1976), that plastids can be transmitted from the male. Indeed, it seems that if plastids are included into the generative cell, they persist into the male germ unit (Dumas, Knox and Gaude, 1985) and are likely to be transmitted into the zygote. Since all sperm cells, of homomorphic or dimorphic type (see Chapter 3), contain mitochondria we must also assume that these organelles are routinely transmitted biparentally.

Ultrastructural modifications to organelles

Information regarding any change which may overcome the genetic constitution of organelle populations after zygote formatuon is not available for higher plants. In *Sacchromycetes,* however, recombination occurs between organelle genomes prior to or during diploidisation. This interaction is swiftly halted as budding commences and the organellar genetic constitution in buds produced by the diploid reverts to one or other of the parental types (Wilkie, 1973). While the nucleus thus appears capable of exerting considerable influence over organelle populations, there would still be selective advan- tages to an organism possessing a system capable of eliminating organelles containing DNA lesions, or simply genomes which are in some way incompatible with the genetic constitution of the new haploid nucleus of the gamete. While a case can thus be made for such a mechanism, improvement of the line must of course be balanced with its survival. For example, were a selection system to operate by the elimination of cells containing 'defective' organellar DNA, and to be present in both male and female cells, it could have a considerable effect on the population size. Were, however, it to operate only in males, where there is debatably a superabundance of gametes, selection could take place with very little cost to the population. It would therefore seem that pollen development would represent a logical starting point to search for such a selection mechanism. Clearly, it would only be of use if heterogenicity exists in organellar genomes within an individual, and, indeed, if potentially dangerous lesions can be shown to occur within organellar DNA. Most modern accounts of the composition of organellar genomes do, describe considerable heterogeneity, and the evidence from plants displaying cytoplasmic male sterility (cms) suggests that the mitochondria of these plants regularly possess lesions in their DNA (Forde and Leaver, 1980). Indeed, plants with cms provide probably the most persuasive evidence for the presence of an organellar selection system operating during gametogenesis for, (at face value in any case,) cells containing organelles with genetic lesions are eliminated - rather than being permitted to form gametes. Further, such a mechanism does not appear to operate in female cells.

In a series of recent investigations, we have followed up earlier classical reports (Guillermond, 1924; Wagner, 1927; Painter, 1943; Sauter and Marquardt, 1967) of unusual and spectacular differentiation occurring during pollen development in flowering plants. It has been possible to demonstrate the presence of a cycle of RNA metabolism (McKenzie, Heslop-Harrison and

Dickinson, 1967; Dickinson and Heslop-Harrison, 1970a; Williams, Heslop-Harrison and Dickinson, 1973) which, whether it is associated with meiosis itself (Huskins and Chen, 1950; P.R. Bell, pers. comm.) or the process of gametogenesis, certainly involves the elimination of most messenger RNA from the microsporocyte. Since it is likely that communication between nucleus and organelles is effected via this type of information carrying molecule, this cycle might in itself expose organelles capable only of survival with "nuclear support". Certainly organelles are affected during this period (Dickinson and Heslop-Harrison, 1970b), and undergo a phase of rapid dedifferentiation followed by a massive synthesis of organellar DNA (Smyth and Shaw, 1979; Bird, Porter and Dickinson, 1983). While this "passive selection" may be important, evidence is now emerging from the Compositae which could be interpreted as indicating that a more active role is assumed by the nucleus in determining the character of the mitochondria transmitted in these plants (Dickinson and Potter, 1979). We report here events occurring in a member of the Compositae which point to a direct interaction taking place between the new haploid nucleus and the population of organelles that is later to be transmitted to the next generation via the sperm cells.

MATERIALS AND METHODS

Techniques used for the fixation of pollen mother cells and microsporocytes of *Cosmos bipinnatus* for light and electron microscopy, are set out in detail in Williams, Heslop-Harrison and Dickinson (1973). Cytochrome C oxidase was localised in the mitochrondria of these cells according to the method of Seligman, Karnovsky, Wasserkrug and Hanker (1968). Data on the numbers of organelles and their situation within the cells were estimated from enlarged micrographs of median sections through single meiocytes or microsporocytes. Estimations were made from at least 10 representative cells from each stage. While the numerical data were analysed statistically with regard to standard error, no attempt was made to quantify the priority of mitochondrial association with the nucleus, since physical attachment was so clear from the micrographs.

Ultrastructural modifications to organelles

RESULTS

Introduction

Although most of the organelles transmitted from one generation to the next are derived from the egg cell, it has been argued in the foregoing that there are good reasons why selection systems should be anticipated in the male as well as female cells. For experimental convenience, therefore, this study has concentrated upon pollen development. Similarly, although mitochondria and plastids are transmitted, the fact that plastids are not transmitted in all species, and also that they are fewer in number, has resulted in the work being concentrated upon mitochondria. Since the developmental stages of interest range from meiotic prophase to the point immediately prior to the break up of the tetrad of microspores, staging of the material is not easy and, of necessity, must be somewhat inaccurate. Whilst meiotic prophase is well defined, the "post meiotic" stage extends from telophase II until the completion of the callose walls delimiting the tetrad, and the "early tetrad" stage ranges from the end of the post meiotic stage to the point at which the sexine of the pollen wall becomes evident.

Meiotic prophase

Prior to meiosis, the mitochondria contained in the archesporial tissue of *Cosmos bipinnatus* are similar to those characteristic of somatic tissue. Little change overcomes these organelles as the cells enter meiosis and (See Fig. 1), at the pachytene stage of prophase, a median section through the meiocyte will reveal the presence of some 50 mitochondria. As meiosis proceeds some spectacular changes occur to these organelles. Firstly they become spherical and rounded, shrinking slightly in maximum dimension, and are often seen to contain little or no internal cristae (see Figs. 2 and 3). A large number of these organelles then accumulate small aggregates of electron-opaque material in the matrix (see Fig. 4). Counts taken during this stage of development indicate that a fall has occurred in the numbers of clearly-identifiable mitochondria (see Fig. 5). The total number of inclusions has, however, increased. As meiotic prophase draws to a close, the numbers of identifiable mitochondria begin to rise again, an increase that is maintained throughout subsequent development (see Fig. 5). Nevertheless, when this increase in mitochondrial number is considered in terms of the accompanying increase the numbers of other inclusions, it may clearly be seen that the

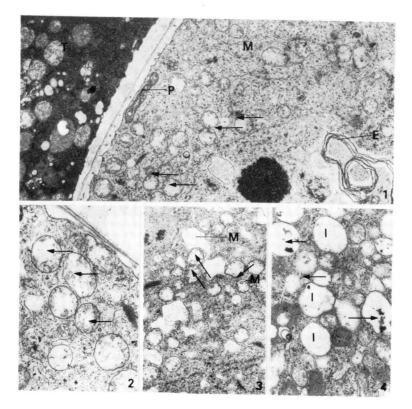

Fig. 1 - Meiocyte (M) and tapetal cell (T) of *Cosmos bipinnatus*
showing "conventional" mitochondria (arrows) and a
plastid (F). The convolutions of the nuclear
envelope (E) characteristic of the onset of meiosis
are clearly visible.   x 12,000

Fig. 2 - Higher power detail of material as shown in Fig. 1.
The mitochondria (arrows) are typical of plant cells.

x 21,800

Fig. 3 - The cytoplasm of a meiocyte entering meiotic prophase.
Some mitochondria (M) are seen to contain central
vesicles (arrows).

X 9,000

Fig. 4 - Cytoplasm of a meiocyte in meiotic prophase. Numerous
spherical inclusions (I) are visible containing
electron opaque material (arrows).

x 25,000

41

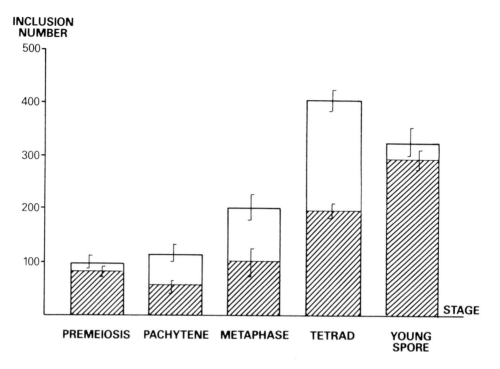

Fig. 5 - Numbers of clearly identifiable mitochondria (hatched portion of
histogram) expressed as a proportion of total spherical inclusion
number (not including plastids). Vertical bars denote standard
errors.

rate of increase of the numbers of total inclusions is much greater than that
of the mitochondria. The mitochondria therefore form a decreasing proportion
of the inclusion population of the cell. As cytokinesis commences a new
inclusion becomes visible in the mitochondria. In many sections this appears
as an internal vesicle bounded by two unit membrane profiles, but careful
sectioning suggests that is in fact an invagination of the mitochondrial
membrane into the matrix of the organelle (see Fig. 6a and 6b). This structure
is possessed by a very large proportion of the mitochondria present at this
and subsequent stages (see Fig 7). The internal vesicles do vary in size
somewhat, but the average dimension is of the order of 200 nm and they must
therefore occupy some 20% of the organellar volume.

Striking changes also overcome the plastids during this period. The normal
amyloplasts present in the archesporial tissue dedifferentiate to form large

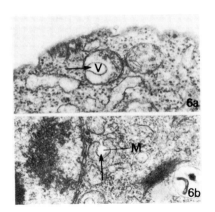

Fig. 6a - Detail of a meiocyte late in meiotic prophase showing
the apparent formation of the central vesicle (V) by
an invagination of the mitochondrial membrane (arrow).

x 34,600

Fig. 6b - Mitochondria (m) displaying central vesicle (arrow),
but present during the tetrad stage.

x 90,000

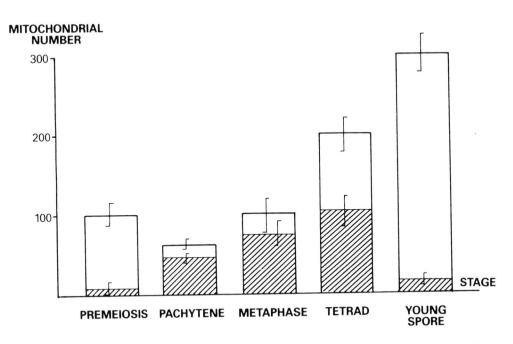

Fig. 7 - Proportion of total mitochondrial population possessing central
vesicles at the various stages of microsporogenesis. Vertical
bars denote standard errors.

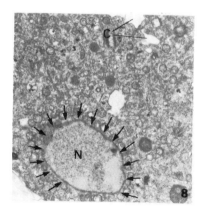

Fig. 8 - A meiocyte of *Cosmos* immediately prior to cytokinesis.
One of the nuclei (N) is visible and is surrounded by
a layer of small mitochondria (arrows). The beginnings
of the cell plates (C) formed between the members of
the tetrad are viable.

pleomorphic structures, staining evenly electron-opaque, and containing very
few internal lamellae. The plastids do not possess the internal vesicles
characteristic of the mitochondria. Interestingly, when observed in the
electron microscope, the meiocytes fall into two populations. The one,
representing the minority of cells, contains meiocytes which possess very
rounded organelles and a particularly electron-lucent cytoplasmic matrix.
The other, in which the cytoplasmic matrix stains evenly electron-opaque,
contains pleomorphic organelles. It is noteworthy that the mitochondria
contained in the latter population contain internal vesicles far more
frequently than those in the former.

Events in the post-meiotic cells
Since both meiosis I and meiosis II in *Cosmos bipinnatus* take place in a
common cytoplasm, telophase II occurs in a large spherical cell bounded by
callose. The newly-formed nuclei are elongated, and come to rest equidistant
from one another (see Fig. 8).

Examination of the cytoplasm at this stage reveals a striking phenomenon
for both counts and straightford observation indicates that a large proportion
of the organelle population of these cells is clustered about the nuclei.

44

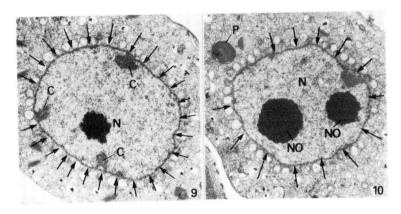

Fig. 9 - Young microspore of *Cosmos* still retained within the
         tetrad and showing a layer of mitochondria (arrows) over
         the surface of the nucleus (n). Chromatin (C) may be
         seen apposed to the point of contact of some mitochondria
         with the envelope.   x 10,500

Fig. 10 - Material as shown in Fig. 9 but featuring the two
          prominent nucleoli (NO) of the nucleus (N),
          mitochondria (arrows) associated with the nuclear
          envelope, and a plastid (P).   x 10,500

Further, closer inspection suggests that a large number of these organelles -
particularly mitochondria - are actually fixed to the nuclear surface
(see Fig. 9-11). While this association between the nuclear envelope and
organelles has been described elsewhere (Dickinson and Potter, 1979) some
details are worth bearing repetition here. Firstly, the adhesion between
organelle and envelope appears to be affected by small rodlets or granules,
some 15-20 nm in maximum dimension (see Figs. 12 and 13). Secondly, serial
sectioning suggests that every mitochondria attached to the envelope in this
fashion appears to carry internal vesicles, with the point of the invagination
never facing towards the envelope. While the staining property of these cells
varies in a manner similar to that seen in prophase meiocytes, examination of
those microsporocytes in which the nuclear material stains well shows that the
point of attachment of each organelle is apposed, on the inner face of the
membrane, by accumulation of electron-opaque chromatin (see Fig. 14).

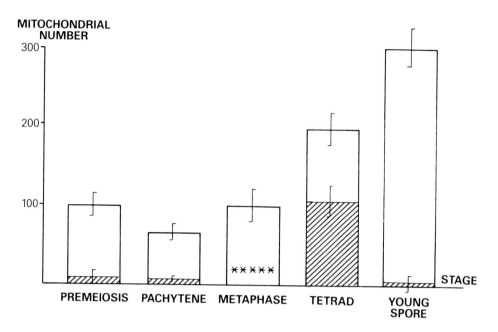

Fig. 11 - Proportion of total mitochondrial population (hatched portion of
histogram) in close proximity to nuclear envelope at the various
stages of microsporogenesis. No estimates made at metaphase since
no organised nucleus present   *****   Vertical bars denote standard
errors.

Most recently, improved fixation techniques have enabled the occasional
detection of sub-structure within this chromatin situated beneath the point o
contact. Here, instead of an aggregation of granules some 20 nm in diameter
there appears to be an ordered invagination of the nuclear envelope, itself
invested by 20 nm granules (see Fig. 15-17). In some cases, these invaginations may travel up to 1 m into the nucleus, where they terminate in small
vesicles which may measure 0.2 m in diameter. Interestingly, these invaginations or tubes do not discharge directly into the cytoplasm, but terminate at
nuclear pore above which an organelle is frequently situated (see Fig. 15).
should be emphasised that these invaginations are not always seen apposing
mitochondria, and not all of them terminate in nuclear pores, but it is clear
that the attachment of the organelles is always reflected by accumulations of
chromatin of some sort. As has been reported earlier (Dickinson and Potter,
1979) attempts to detach organelles from the nuclear envelope by inflating th

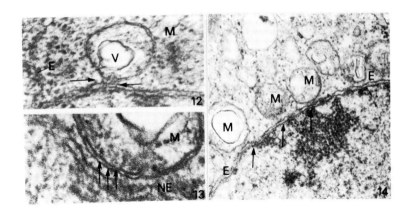

Fig. 12 - Detail of material at a stage similar to that
shown in Fig. 10, but at a higher magnification. A
mitochondrion (M) containing a central vesicle (V)
is seen attached to the nuclear envelope (E). Small
granules (arrows) appear to attach the organelle to
the nuclear envelope.

x 71,700

Fig. 13 - Material as shown in Fig. 12, but showing small rodlets
(arrows) apparently attaching the mitochondrion (M) to
the envelope (E).

x 270,000

Fig. 14 - Mitochondria (M) apparently attached to the nuclear
envelope (E). Elements of the chromatin (arrows)
appear to be apposed to the point of contact between
the organelle and the envelope.

x 53,000

cells osmotically have proved fruitless. We have also exposed microsporocytes
at this stage to a nitrogen atmosphere of periods between 30 minutes and 120
minutes, with no effect to the situation of the organelles. Experiments
involving microsporocytes treated with dinitrophenol (DNP) provided somewhat
equivocal results, but the mitochondria contained in the vast majority of
these cells remained affixed to the nucleus.

As is clear from Fig. 11, not all the mitochondria present in the tetrad
are attached to the nuclear envelope . These other organelles are, however,

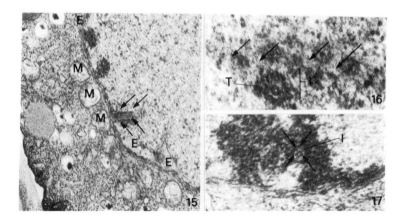

Fig.15 - Further details of the points of attachment between
mitochondria (M) and nuclear envelope (E) but, at one
point (arrows) a nuclear invagination is visible.

x 30,000

Fig. 16 - Tangential section of the nuclear surface of a tetrad
nucleus showing nuclear pores (arrows) and the beginnings
of a nuclear invagination (I), Note the central tubule (T).

x 70,000

Fig. 17 - Transverse section of a nuclear invagination (I)
invested by darkly staining chromatin (arrows).

x 80,000

restricted to a fairly thin layer measuring about $5\mu m$ in depth, at the cell
periphery. The region between this layer and the mitochondria investing the
nucleus is occupied by large numbers of vesicles ranging in size between
0.25 and $0.75\mu m$ in diameter. Since there are far fewer plastids than
mitochondria in this cytoplasm, it is less easy to follow the behaviour of
these organelles. This task is made even more difficult by the fact that
plastids are extremely pleomorphic and may extend for several μm through
the cytoplasm. Although in section very few plastids would appear to make
contact with the nuclear envelope, serial sectioning suggests that some do
(see Fig. 18). Both populations of mitochondria, that at the cell periphery
and that associated with the nucleus, contain cytochrome C oxidase (see Fig.

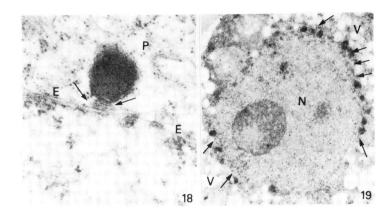

Fig. 18 - Plastid (P) apparently attached to the envelope (E) of
a tetrad nucleus (N) by fibrillar material (arrows).

x 53,000

Fig. 19 - Material at the tetrad stage of development reacted
to reveal the presence of cytochrome C oxidase.
Reaction product is deposited within the mitochondria
(arrows) investing the nucleus (N). The population
of vesicles (V) contain little or no reaction product.
Control preparations which had been subject to boiling
for 5 mins, or had been treated with KCN, gave no
such reaction product.

x 13,5000

The amounts of this enzyme possessed by these organelles appears to be less
than that contained in tapetal mitochondria, but some level of activity may be
detected in every clearly-identifiable microsporocyte mitochondrion (see Figs.
20 and 21). Interestingly, cytochrome C oxidase does not seem to be present
on the internal vesicle possessed by many of these organelles (see Fig. 21).

The tetrad stage
As the microsporocyte nuclei become rounded and the first stages of sexine
formation take place, fewer organelles appear to be associated with the nuclear
envelope. Instead, spherical mitochondria containing internal vesicles are
once again seen in large numbers distributed throughout the cytoplasm (see

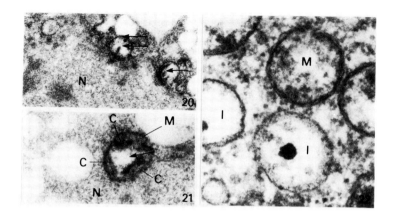

Fig. 20 - Higher power study of material as shown in Fig. 19. The
reaction product is shown to be associated with
mitochondrial cristae (arrows). The nucleus (N) appears
particularly diffuse. Control preparations gave the
same result as in Fig. 19.

x 30,000

Fig. 21 - Single mitochondria (M) at the surface of a tetrad
nucleus (N) reacted to show the presence of cytochrome
C oxidase. The enzyme can be seen associated with the
cristae (C) but not with the central vesicle (arrow).
Controls as in Figs. 19 and 20.

x 70,000

Fig. 22 - Range of structures present in the late-tetrad
cytoplasm. These range from inclusion resembling
mitochondria (M), to those possessing a single bounding
membrane and containing electron opaque deposits (I).

x 105,000

Fig. 22). The striking stratification characteristic of the immediate post-
meiotic stage is lost, and mitochondria now begin to reassume their normal
"somatic" morphology (see Fig. 23). The organelles enlarge, lose any internal
vesicle, and some assume a more elongate form. Interestingly, a residual
population of small spherical organelles survives (see Fig. 24) frequently
containing masses of electron-opaque material. These decrease in number until
by the stage the young sexine of the pollen wall is fully defined, they are
absent altogether. As with the mitochondria, the plastids move away from the

nucleus during the tetrad stage, and become dispersed evenly within the
cytoplasm. They also assume a more rounded morphology and, in some cells,
begin to accumulate starch.

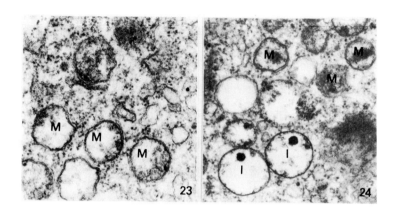

Fig. 23 - Clearly identifiable mitochondria (M) visible in
the mid-tetrad cytoplasm.

× 36,000

Fig. 24 - Cytoplasm of the late tetrad showing the presence
of a few remaining inclusions (I) containing electron
opaque deposits, and some structures resembling
mitochondria (M) possessing diffuse electron opaque
material.

x 43,800

DISCUSSION

Organelle population dynamics during meiosis and microsporogenesis
Although some data are available on the numbers of organelles present during
microsporogenesis in angiosperms (Willemse, 1972), little is known of the
dynamics of the various organelle populations. A major complication is provided
by the range of organelle structure observed, extending between clearly-identifiable mitochondria to simple spherical "necrotic" vesicles. Even cursory
inspection of the micrographs reveals that all these inclusions are
structurally related, and all stages in the conversion of one form to the
other may be observed. The events occurring with the "mitochondrial" population

51

must therefore be considered in terms of changes taking place in this far larger population of inclusions. Since the range of morphologies observed has no obvious explanation except in that it represents the progressive necrosis o mitochondria, the pattern of development during meiotic prophase as set out in Fig. 5 requires some reinterpretation. During meiotic prophase, therefore, a substantial increase must occur in the numbers of mitochondria present in the meiocytes and, as this population increases in size, an increasing proportion o it fails to develop normally, forming instead necrotic vesicles. Thus some 50% of the total "mitochondrial" population is necrotic by the tetrad stage. This proliferation of organelles ceases following the formation of the haploid microspores, and it would appear that the necrotic organelles are finally degraded and absorbed into the tetrad cytoplasm.

Certainly both mitochondria and plastids increase dramatically in number during meiotic prophase in other species (e.g. *Lilium,* Dickinson and Heslop-Harrison, 1970b), and it has recently been shown that this increase is accom-panied by a very rapid synthesis of organellar DNA (Smyth and Shaw, 1977; Bird, Porter and Dickinson, 1983). Owing to their extreme pleomorphism, it has not been possible to examine the population dynamics of plastids during microsporogenesis in *Cosmos,* but simple counts from micrographs suggests that they too increase in number.

During the tetrad stage over 50% of the mitochondria present in the cell become associated with the nuclear envelope and, since the generation of organelles, particularly mitochondria, from the nuclear envelope has been previously reported in the literature (Bell and Muhlethaler, 1964; Bell, Frey Wyssling and Muhlethaler, 1966) the possibility that such an event occurs here should be examined. However, the fact that the major increase in organelle number seems to occur during meiotic prophase, long before any organelles beco associated with the nucleus, would suggest that it is unlikely that mitochondr are being formed *de novo* from the nuclear envelope, whatever the nature of the association between the organelles and the nucleus.

Changes in mitochondrial structure and situation

The conversion of mitochondria from the normal 'somatic' form to isodiametric organelle in early meiotic prophase is well documented (Bal and De, 1961; Dickinson and Heslop-Harrison, 1970b). The small rounded organelles are terme "promitochondria" and seem to be capable of very rapid division. This phase of division is accompanied by high levels of DNA synthesis (Bird, Porter and Dickinson, 1983), and preliminary studies suggest that these organelles contain comparatively small amounts of cytochrome C oxidase. The appearance of the central vesicle possessed by many of the organelles in prophase is difficult to explain. This structure appears to be formed by a simple invagination of both the mitochondrial membranes, and clue to its origin may lie in the two forms of cells seen in the electron micrographs. One is generally electron-opaque and contains mitochondria with vesicles, while the other is far more electron-lucent, and contains few organelles containing these structures. Since the darkly-staining cells also appear to be shrunk in from the callose walls, it may be that they are plasmolysed as a result of some form of osmotic stress. If this is the case a similar osmotic factor could affect the organelles of the cells, causing a weak region of the organelle membrane to become distended to form the vesicles observed. The lack of cytochrome C on the vesicle would support such an inference. However, since the vesicles are so regular, and since they are occasionally also seen in the more lightly-staining cells would suggest that, whether or not they are an artefact, they are caused by a stage-specific modification of the mitochondrial membranes.

By the time the tetrad stage is reached viable mitochondria appear to be restricted to two principle domains, a peripheral layer just beneath the cell surface, and a layer appressed to the surface of the nuclear envelope. Structural and cytochemical analyses suggest that the vesicles occupying the region between these two regions are unlikely to fill any mitochondrial function. The association between mitochondria and the nuclear envelope in *Cosmos* is discussed at length in another publication (Dickinson and Potter, 1979), but information emerging from this study may throw some light on this unusual interaction. The earlier work indicated that two alternatives only could account for the association between mitochondria and the nuclear envelope at this stage of development; either these organelles were providing energy in the form of ATP to the new haploid nucleus, or they were receiving information-carrying, or other, molecules from the nucleus. Certainly a situation could arise by which organelles aggregated around the nuclear envelope as a result of changes occurring to vesicle populations elsewhere in the cell, but the fact that chromatin is always observed associated with the point of contact, combined with the observation that efforts to detach the organelles from the nuclear surface have failed (Dickinson and Potter,

1979) would suggest that the association is of a more positive nature. Further the treatment of isolated tetrads with Dinitrophenol (DNP), and high levels of nitrogen, does not result in their detachment from the nucleus which indicates that if their role is to provide energy, the cytoplasmic apparatus for attaching them to the nuclear surface must be sophisticated and unrelated to their production of ATP. Perhaps most interesting is the new observation that the areas of chromatin apposed to the contact-point of the mitochondria may sometimes have some substructure. The type of nuclear invagination observed is identical with that seen in the gymnosperms (Dickinson and Bell, 1972; Dickinson and Potter, 1975). In *Pinus banksiana*, for example, it has been shown that these invaginations are formed from a modified nuclear pore, and that they consist of a channel extending into nucleus, formed from nuclear envelope, and invested by regions of chromatin known to be rich in DNA. The content of the invagination, and particularly that of its terminal vesicle has yet to be positively identified in *Pinus*, but certainly contains high levels of RNA. These structures are interpreted as representing a mechanism by which information-carrying molecules, presumably messenger-RNA, is exported rapidly from the new haploid nucleus. From our present data it is impossible to determine whether the invaginations in *Cosmos* are always situated at the point of contact with the mitochondria, or simply that they are generally distributed over the nuclear surface and fortuitiously coincide with an organellar attachment area.

Redifferentiation of the organelle population takes place late in the tetrad stage following the detachment of the mitochondria from the nuclear envelope. From the methods employed in this study it is impossible to determine whether all the mitochondria present have spent a period in association with the nuclear envelope. Certainly they all appear identical and the numbers associated with the nucleus at any one time would suggest that contact by all the organelles would be feasible. The precise pathway of redifferentiation is identical in organelles in both "nuclear" and "peripheral domains, and is described in Dickinson and Potter, (1979). The young microspore, as it is released from the tetrad, thus contains a population of mitochondria derived entirely from those organelles previously either associated with the nucleus, or sited at the periphery of the cell.

The significance of organellar behaviour during microsporogenesis

The principal features of mitochondrial differentiation during microsporogenesis in *Cosmos* would seem to consist of (i) a cycle featuring dedifferentiation, rapid replication accompanied by a high rate of mortality, and
subsequent redifferentiation to normal organelles, and (ii) a period of
association between clearly recognisable mitochondria and the nuclear
envelope.

The cycle of organellar differentiation and elimination should however,
be regarded from two viewpoints. Firstly, the physiological reasons for
the organellar dedifferentiation, replication and elimination should be
examined but, equally importantly, the effect that these events have on the
subsequent fate of organelles in the next generation should also be considered.
The 'communication links' between the nucleus and organelle populations have
been proposed to undergo dramatic changes during meiosis (Porter, Bird
and Dickinson, 1983), for levels of nuclear-encoded manager RNA fall to barely-
detectable levels over this period and it is reasonable to presume that much
organellar activity is controlled by means of this species of nucleic acid.
Clearly the mitochondrial population cannot completely lose touch with the
nucleus, for nuclear participation is required for most mitochondrial
functions, and certainly for particular steps in the replication of these
organelles. Nevertheless, it is not beyond the bounds of possibility that
a decrease in nuclear participation in organellar control may result in the
dedifferentiation of these organelles and, perhaps, their uncontrolled
replication. Further, it is possible that during the somatic life of an
individual the nucleus so controls activities of the organellar population
that regions of defective organellar DNA are not transcribed, and even
that the presence of defective genes in the mitochondrial DNA may be
'compensated' by the expression of nuclear genes. The lack of nuclear
participation characteristic of meiosis would thus therefore serve to
decrease the stringency of this control, permitting organelles to synthesise
potentially harmful compounds and, more speculatitively, leave defective
organelles without complementation from nuclear genes.

Whatever the details of the nuclear organellar interactions taking place
at this time, the fact remains that of the mitochondria present in the
Cosmos meiocyte at the end of prophase, only 50% survive to be included into
the young microspore. Some selection mechanism must clearly be at work at

55

this stage, with the fittest organelles surviving for transmission. Whether or not the selection involves the identification and elimination of lesions in mitochondrial DNA, or simply competition for nuclear encoded mitochondria membrane components by individual organelles, remains to be determined. Megasporogenesis has yet to be studied in *Cosmos*, and we are therefore ignorant as to whether parallel changes occur in the development of the female cells. Certainly, in *Lilium,* inclusions resembling necrotic mitochondria are visible in the prophase cytoplasm (Dickinson and Potter, 1978) but measurements have not been made of the organellar populations.

It is tempting to identify the phase of mitochondrial elimination seen during meiotic prophase with the operation of certain cms systems. We have made a preliminary study of cms in *Helianthus,* which is closely related to *Cosmos*, and have found that degeneration of the reproductive tissue takes place around the tetrad stage of development. It remains quite feasible therefore that in some cms systems lesions in mitochondrial DNA are so comprehensive that the stress put upon the organelle population during meiotic prophase is such that not 50%, but 100% are eliminated. If this were the case, necrosis of the reproductive cells would surely follow. Clearly, such an event can not explain all types of cms, or, even all types of cms observed in *Helianthus* (Horner, 1977). There is certainly a great contrast between the position in *Cosmos* where 50% of organelles are eliminate but the cells survive, and cms lines when 100% of the cells abort. It should be noted , however, that our present cms lines are the result of repeated selection by plant breeders, and lines in which only incomplete cms occurred would have been discarded. If indeed parallel events do occur in female cells the fact that cytoplasmic female sterility does not exist must be explained. The megaspore and microspore differ quite comprehensively in their situation, in that the microspore is isolated in a wall of callose, and floats detached in the loculus whereas the megaspore is bounded by thin, relatively permeable, wall and is firmly embedded in the nucellar tissue. Female cells would thus be well cushioned to survive elimination of very high proportions of their organelles.

The association of mitochondria with the nuclear envelope would seem unlikely to be associated with the cycle of differentiation and elimination described above. However, our observations suggest that this association is unlikely simply to involve the transfer of energy from active mitochondria

into the new haploid nucleus. The discovery that the chromatin subjacent
to the mitochondria is sometimes organised in the form of nuclear invaginations,
commonly regarded to be a sign that information carrying molecules are being
exported from the nucleus (Aldrich and Vasil, 1970; Dickinson and Bell,
1972; Dickinson and Potter, 1975) makes it tempting to suggest that such
molecules are being passed to the organelles. No instances of physical contact
between organelles and the nucleus leading to transfer of known macromolecules
have been recorded, but mitochondria have been reported to be associated with
the nuclear envelope in a number of organisms ranging from Man (Baker and
Franchi, 1969) to *Myosurus minimus* (Woodcock and Bell, 1968). Since there
is much evidence, derived principally from the work on *Saccharomyces* (Wilkie,
1973), that the nucleus is capable of selecting "appropriate" organellar
genotypes, it is not beyond the bounds of possibility that in *Cosmos,* the
new haploid nucleus 'conditions' the old diplophase organelles in an
analogous manner.

ACKNOWLEDGEMENTS

This work was supported in part by a grant from the Agriculture and Food
Research Council of the UK. The author's thanks are due to Clare Willson
for valuable technical assistance, to Nichola Slee for help with the
stereological investigations, and to Susan Mitchell for drawing the figures.

REFERENCES

Aldrich, H.C. and Vasil, I.K. (1970). Ultrastructure of the post-
    meiotic nuclear envelope in microspores of *Podocarpa macrophylia.*
    J. Ultrastr. Res. 32, 307-315.
Baker, T. and Franchi, L.K. (1969). The origin of cytoplasmic inclusions
    from the nuclear envelope of mammalian oocytes. *Zeit. fur Zellforsch.*
    93, 45-55.
Bal, A.K. and De, D.N. (1961). Developmental changes in the sub-microscopic
    morphology of cytoplasmic components during microsporogenesis in
    Tradescantia. Dev. Biol. 3, 341-54.
Bell, P.R. and Muhlethaler, K. (1964). The degeneration and reappearance
    of mitochondria in the egg cells of a plant. *J.Cell Biol. 20,* 235-48.

Ultrastructural modifications to organelles

Bell, P.R., Frey-Wyssling, A. and Muhlethaler, K. (1966). Evidence for the discontinuity of plastids in the sexual reproduction of a plant. *J. Ultrastr. Res. 15,* 108-21.

Bird, J., Porter, E.K. and Dickinson, E.K. (1983). Events in the cytoplasm during male meiosis in *Lilium. J. Cell Sci. 59,* 27-42.

Dickinson, H.G. and Bell, P.R. (1972). Structures resembling nuclear pores at the orifice of nuclear invaginations in developing microspores of *Pinus banksiana. Dev, Biol. 27,* 425-429.

Dickinson, H.G. and Heslop-Harrison, J. (1970a). The ribosome cycle, nucleoli and cytoplasmic nucleoloids in the meiocytes of *Lilium. Protoplasma 69,* 187-200.

Dickinson, H.G. and Heslop-Harrison, J. (1970b). The behavious of plastids during meiosis in the microsporocyte of *Lilium longiflorum* Thunb. *Cytobios. 6,* 103-118.

Dickinson, H.G. and Potter, U., (1975). Post-meiotic nucleo-cytoplasmic interaction in *Pinus banksiana:* the secretion of RNA by the nucleus. *Planta (Berl.) 122,* 99-104.

Dickinson, H.G., and Potter, U.(1978). Cytoplasmic changes accompanying female meiosis in *Lilium longiflorum* Thunb. *J. Cell Sci. 29,* 147-169.

Dickinson, H.G., and Potter, U. (1979). Post-meiotic nucleo-cytoplasmic interaction in *Cosmos bipinnatus. Planta 145,* 449-457.

Dumas, C., Knox, R.B. and Gaude, T. (1985). The mature viable tricellular pollen grain of *Brassica:* germ line characteristics. *Protoplasma* (details not available).

Forde, B.G. and Leaver, C.J. (1980). Nuclear and cytoplasmic genes controllin synthesis of variant mitochondrial polypeptides in male-sterile maize. *Proc. Natl. Acad. Sci. US,* 77, 418-422.

Guillermond, A. (1924). Recherches sur l'evolution du chondriome pendant le developpement du sac embryonnaire et des cellules meres des grains de pollen dans les Liliaceases et sur la signification des formations ergastoplasmiques. *Ann. Sci. Nat. Bot. 6,* 1-52.

Hagemann, R. (1976) Plastid distribution and plastid competition in higher plants and the induction of plastome mutations by Nitroso-urea compounds. In *Genetics and Biogenesis of Chloroplasts and Mitochondria*. (eds. Bucher, F., *et al*.), Amsterdam, North Holland. 331-338.

Horner, H.T. (1977). A comparative light- and electron-microscopic study of microsporogenesis in male fertile and cytoplasmic male sterile sunflower (*Helianthus annus*). *Amer. J. Bot. 64*, 745-759.

Huskins, C.L., and Chen, K.L. (1954). Segregation and reduction in somatic tissue. *J. Hered. 41*, 13-18.

Mackenzie, A., Heslop-Harrison, J. and Dickinson, H.G. (1967). Elimination of ribosomes during meiotic prophase. *Nature 215,* 997-999.

Painter, T.S. (1943). Cell growth and nucleic acids in the pollen of *Rhoeo discolor*. *Bot. Gaz. 105,* 58-68.

Russell, S.D. (1986). Dimorphic sperm cells, cytoplasmic transmission and preferential fertilisation in the synergidless angiosperm, *Plumbago zeylanica*. (Chapter 4, this volume).

Sauter, J.J. and Marquardt, H. (1967). Die Rolle die Nukleohistons bei der RNS-und Protein synthese wahrende der Mikrosporogenese von *Petunia tenuifolia* L. Zeit. *Pflanzenphysiol. 58,* 126-137.

Seligman, A.M., Karnovsky, M.J., Wasserkrug, H.L. and Hanker, J.S. (1968) Non-droplet ultrastructural demonstration of cytochrome oxidase activity with a polymerising osmiophilic reagent, diaminobenzidine (DAB). *J. Cell Biol. 38,* 1-14.

Smyth, D.R. and Shaw, T.S. (1979). Cytoplasmic DNA synthesis at meiotic prophase in *Lilium henryi*. *Aust. J. Bot. 27,* 273-284.

Wagner, N. (1927). Evolution du chondriome pendent la formation de pollen des angiosperms. *Biolog. gen. 3,* 15-66.

Wilkie, D. (1973). Cytoplasmic genetic systems of eukaryotic cells. *British Medical Bulletin 29,* 263-68.

Willemse, M. (1972). Morphological and quantitative changes in the population of cell organelles during microsporogenesis of *Gasteria verrucosa*. *Acta. Bot. Neerl. 21,* 17-31.

Williams, E.G., Heslop-Harrison, J. and Dickinson, H.G. (1973). The activity of the nucleolus organising region and the origin of cytoplasmic nucleoloids in meiocytes of *Lilium*. *Protoplasma 77,* 79-93.

Ultrastructural modifications to organelles

Woodcock, C.L.F. and Bell, P.R. (1968). Features of the ultra-structure of the female gametophyte of *Myosurus minimus*. *J. Ultrastr. Res. 22,* 546-563.

3 Mitochondrial delivery via the male gametophyte and the prospects for recombination

G. P. Chapman

ABSTRACT   A feature of plants to emerge recently is the size and complexity of their mitochondrial genomes. Since sexual fusion creates the basis of nuclear diversity, evidence for the basis of organellar diversity originating from both sexes entering the zygote, is examined here for cryptogams and phanerogams.

A common pattern to be found is the persistence of the mitochondria of one sex and 'disintegration' of the other (not necessarily that of the male). Arguably, anisogamy conserves mitochondrial genomes but with an intermittent or small input from (usually) the smaller gamete making mitochondrial recombination possible as part of the adaptive process under the influence of the new zygote nucleus.

Although some crop plant mitochondrial genomes are among the most closely analysed, the possibility of mt-DNA of male origin contributing to mitochondrial diversity, still requires both cytological and molecular examination.

INTRODUCTION

Since protoplast fusion is the only known means to induce mitochondrial recombination (Hanson and Conde, 1985) and, arguably, gamete union is a specialised form of protoplast fusion (van Vent and Willemse, 1984), what prospects are there for mitochondrial recombination in the zygote? Is it possible that anisogamy could conserve mitochondrial genomes but with an intermittent input from the smaller gamete as part of the adaptive process? Among flowering plants, the likelihood of mitochondrial transmission from the male gametophyte probably varies with the genus but before considering this, mention will be made of more primitive plants.

Ultrastructural modifications to organelles

NON-FLOWERING PLANTS

The most complete evidence is from yeast where Fonty, Goursot, Wilkie and
Bernardi (1978) showed for *Saccharomyces cerevisiae*, recombinant mitochondrial
genomes arising from what were interpreted as unequal crossover events. For
other organisms evidence is more circumstantial. Moestrup (1975) noted that
for eukaryotic algae, apart from the nucleus, only mitochondria occurred
in all gametes. Hoffman (1974) described in *Oedogonium* persistent mitochondri
of male origin near the zygote nucleus. Duckett (1975) drew attention to the
relation between numbers and sizes of mitochondria and the cell size and
numbers of flagella of spermatozoids. Thus, bryophytes (biflagellate) have
one or two mitochondria. In leptosporangiate ferns there are 50 to 60
mitochondria. In *Equisetum* with larger spermatozoids and more flagella
there are several hundred mitochondria. Cycads with more than 10,000
flagella have several thousand mitochondria. An obvious conclusion is that
the energy demands of mobility explain this range of mitochondrial numbers
but this may be an oversimplification when considering the following examples.

For *Marsilea vestita,* Myles (1974) reported that mitochondria were mostly
associated with the cytoplasmic vesicle shed before sperm entry but that
one mitochondrion, asscciated with the male nucleus was delivered to the egg
cytoplasm and that its fate could be followed. Since this mitochondrion was
destroyed within a few hours of entry, Myles (1978) concluded that it was
unlikely to play any genetic role. The argument however, could be inverted.
Only if the male mitochondrion were rapidly 'disintegrated' so that
mitochondrial recombination preceded the first division could the male genome
be represented in all descendant cells. By extension, if in the zygote
only some mitochondria were changed, new and old types might not be equally
represented in subsequent cells.

In the gymnosperm *Cephalotaxus drupacea*, two male nuclei enter the egg
although only one fuses with the egg nucleus. A quantity of male cytoplasm
rich in mitochondria, plastids and ribosomes also enters the egg at gamete
entry. Prior to this, the egg chondriome apparently degenerates and
Gianordoli (1974) considered that, in this particular case, all of the
plastids and mitochondria and most of the ribosomal RNA are of male origin.

According to Chesney and Thomas (1971), in *Pinus nigra* male mitochondria
are delivered to the egg, the mitochondria in the latter showing 'structural

regression'. The same situation applies for male gametes of *Biota orientalis* (Cupressaceae) and only a few mitochondria of egg origin are said to persist in the zygote.

Common to the fern and gymnosperm examples is the preferential survival of the mitochondrial genome of one sex with an uncertain role for the other that 'distintegrates' and this, according to species, can be of either sex.

In the trend from isogamy to anisogamy, the nuclear contributions remain about equal but the plastid either 'disintegrates' on arrival or is isolated before sperm entry. Mitochondria seem intermediate in this respect and at least, a proportion of the mitochondrial genome from the small gamete, appears generally to reach the egg.

FLOWERING PLANTS

Pollen

Dickinson and Heslop-Harrison (1977) have described how, in meiosis, a remnant of both plastid and mitochondrial structure is conserved and redifferentiated.

In angiosperms, meiosis benefits the sporophyte primarily, when seen as a chromosome event. Could it be that where the chondriome is ostensibly 'reworked' at meiosis, survives synergid passage and enters the zygote and that, again, the subsequent sporophyte, and not the transient male gametophyte, is the principal beneficiary?

Uninucleate pollen divides to give a vegetative and a generative cell. Plastids often associate with the former but mitochondria occur at least initially in both. Table I sets out four alternative developmental patterns. Surprisingly, *Solanum* apparently contrasts with *Lycopersicon* and for the former Clauhs and Grun (1977) report persistent mitochondria in the generative cell during tube growth in the style. Wilms and Keijzer (1985) have shbwn, for spinach, joined male cells each with a mitochondrion. Russell and Cass (1983) using *Plumbago zeylanica* showed that plastid-laden male cells, perhaps predestined to enter the egg, still contain more than 20 mitochondria.

Ultrastructural modifications to organelles

Table I* Contrasted Modes of Organelle Transmission from Pollen

Stage of Development e.g.	Type I Lycopersicon	Type II Antirrhinum	Type III Pelargonium	Type IV Plumbago
Uninucleate microspore	Plastids and mitochondria both present			
1st Division vegetative and generative cells	Excluded from g.c.	PLASTIDS Often excluded from g.c.	Present in g.c.	Present in g.c.
	present but then degenerate in g.c.	MITOCHONDRIA persistent in g.c.	persistent in g.c.	persistent in g.c
2nd Division generative cell gives two male cells	--	PLASTIDS sometimes enter m.c.'s	often enter m.c.'s	preferentially enter one m.c. (to egg)
	remnants only ?	MITOCHONDRIA present in m.c.'s	present in m.c.'s	preferentially enter alternative m.c. (to central cell)
Synergid passage	-	PLASTIDS sometimes survive	often survive	No Synergid
	seldom survive	MITOCHONDRIA sometimes survive	often survive	
Within zygote	-	PLASTIDS sometimes survive	often survive	apparently persist
	?	MITOCHONDRIA ?	often survive	if present apparently persist

g.c. generative cell
m.c. male cell
v.c. vegetative cell
* Compiled from data and observations of Hagemann (1979) and Russell (1985)

Synergids

Although the cause and timing of synergid collapse varies among species, it is briefly, four nucleate when together with its own, all three nuclei from the pollen tube are present. Modification of the male cell within the synergid is obscure but could involve removal of some or all contents except the nucleus. If the male cell were sufficiently intact to permit protoplast fusion *sensu strictu* then some or all of the chondriome (presumably between the outer nuclear and the plasmalemma membranes) could be incorporated into the zygote cytoplasm. Hagemann (1979) commented that for *Oenothera, Pelargonium* and *Hypericum*, each of which possess biparental plastid inheritance, paternal mitochondria were also transmitted to the zygote. In an electron microscope study of plastid inheritance in *Oenothera erythrosepta* Meyer and Stubbs (1974) were able to identify plastids by their parentage but evidently could not extend this to mitochondria. It does seem however, that in some cases both mitochondria and plastids seem to survive synergid passage and enter the zygote although much more data is desirable.

The megagametophyte

Russell (1983) found in *Plumbago* (a genus lacking synergids) that organelles from the vegetative cell and its nucleus were contained between the egg and central cells but that organelles within the male gametes were preferentially absorbed.

Mitochondrial diversity in Type 1 organisms

In Type 1 there is, on the available cytological evidence, least likelihood of mitochondrial transmission from male gametophyte to zygote cytoplasm. Both *Triticum* and *Brassica* are examples of Type 1 and, for each, mitochondrial heterogeneity has been described (for wheat; Falconet, Lejeune, Quetier and Gray, 1984; and for cauliflower; Chetrit, Mathieu, Muller and Vedel, 1984). Clearly therefore, an hypothesis to explain the generation of such diversity is now needed.

A MECHANISM

'Transfer' can be regarded as genetic exchange between different kinds of organelle and 'recombination' an exchange between similar kinds and is thus a 'non-promiscuous' event. This latter conforms to the interpretation of Belliard, Vedel and Pelletier (1979) for mitochondrial recombination following protoplast fusion in *Nicotiana*.

Within the zygote the following process is envisaged: DNA, of immediately mitochondrial origin only, is involved. The egg mitochondrial genome could be *'homo-' or 'heteroplastic', i.e. consisting of one or more kinds. Together with the male nucleus, there could be transferred particles varying from relatively minute fragments of mt-DNA to whole mitochondria, depending on species. At fertilisation, the zygote becomes mitochondrially heteroplastic if it were not so already and then, depending on the overriding influence of the new zygote nucleus, the original egg mitochondria or those modified by recombination multiply preferentially in subsequent cell generations.

If events such as these were to happen, it would help explain why 'reworking' the chondriome at meiosis coincides precisely with the derivation of recombinant nuclei. It is suggested that such reworking is of only relictual significance if the male chondriome does not survive to enter the zygote but if it does then each haploid nuclear 'sample' of the diploid pollen mother nucleus requires the appropriate retinue of suitably constituted mitochondria.

Such a mechanism might operate effectively in every zygote or much more rarely. It would not explain how the mitochondrial genome of plants is so complex but it would be a basis for such complexity to re-arrange. And is it not the case that among sexually reproducing organisms the smaller the mitochondrial genome the less likely is it to be biparentally transmitted?

REFERENCES

Birky, C.W., Acton, A.R., Dietrich, R. and Carver, M. (1982). Mitochondrial transmission genetics: Replication, Recombination and Segregation of Mitochondrial DNA and its inheritance in **crosses**. 337 - 348. In *Mitochondrial Genes*. (eds. Slonimski, P., Bort, P. and Attardi, G.) pusb. Coldspring Harbour, 500.

Chesney, L. and Thomas, M.J. (1971). Electron microscopy studies in gametogenesis and fertilisation. *Phytomorphology 21*, 50 - 63.

---

*As defined by Birky, Acton, Dietrich and Carver (1982)

Chetrit, P., Mathieu, C., Muller, J.P. and Vedel, F. (1984a). Physical
and gene mapping of cauliflower (*Brassica oleracea*) mitochondrial DNA.
Current Genet. 8, 413 - 421.

Clauhs, R.P. and Grun, P. (1977). Changes in plastid and mitochondrial
content during maturation of generative cells of *Solanum* (Solanaceae).
Amer. J. Bot. 64, 377 - 383.

Dickinson, H.G. and Heslop-Harrison, J. (1977). Ribosomes, membranes and
organelles during meiosis in angiosperms. *Phil. Trans R. Soc. London
277,* 327 - 342.

Duckett, J.G. (1975). Spermatogenesis in pteridophytes 97 - 127. In
The Biology of the Male Gamete. (eds. Duckett, J.G. and Racey, P.A.)
Academic Press, London, 460.

Falconet, D., Lejeune, B., Quetier, F. and Gray, M.W. (1984). Evidence for
homologous recombination between repeated sequences containing 18S and
5S ribosomal RNA genes in wheat mitochondrial DNA. *EMBO J. 3,* 297 - 302.

Fonty, G., Goursot, R., Wilkie, D. and Bernardi, G. (1978). The mitochondrial genome of wild type yeast cells vii. Recombination in crosses.
J. Mol. Biol. 119, 213 - 235.

Gianordoli, M, (1974). A cytological examination on gametes and fecundation
among *Cephalotaxis drupacea*. In *Fertilisation in Higher Plants*. (ed.
Linskens, H.F.) North Holland Pub. Amsterdam 373.

Hagemann, R. (1979). Genetics and molecular biology of plastids of higher
plants. *Stadler Symp. 11,* 91 - 116.

Hanson, M.R. and Conde, M.F. (1985). Functioning and variation of cytoplasmic genomes: Lessons from cytoplasmic-nuclear interactions affecting
male fertility in plants. *Int. Rev. Cytol. 94,* 213 - 267.

Hoffman, L.R. (1974). Fertilisation in *Oedogonium*. II Karyogamy. Amer.
J. Bot., 61, 1067 - 1090.

Leaver, C.J. and Gray, M.W. (1982). Mitochondrial genome organisation and
expression in higher plants. *Ann. Rev. Plant Physiol. 33,* 375-402.

Meyer, B. and Stubbs, W. (1974). Das Zahlenverhaltuis von mutterlichen und
vaterlichen plastiden in der zygoten von *Oenothera erythrosepala,* Bortas
(syn *Oe. lamarkiana*). Berichtung der deutschenbotanischen Gesellshaflt. *87,*
29 - 38.

Moestrup, O. (1975). Some aspects of sexual reproduction in eukaryotic algae.
pp. 23-37 In *The Biology of the Male Gamete.* (eds. Duckett, S.G. and

Racey, P.A.) Academic Press, London. pp.460.

Myles, D.G. (1975). Structural changes in the sperm of *Marsilea vestita* before and after fertilisation. pp. 129-134 In *The Biology of the Male Gamete*. (eds. Duckett, J.G. and Racey, P.A.) Academic Press, London pp.460.

Myles, D.G. (1978). The fine structure of fertilisation in the fern *Marsilea vestita*. *J. Cell Sol. 30,* 265-281.

Russell, S. and Cass, D. (1983). Unequal distribution of plastids and mitochondria during sperm cell formation in *Plumbago zeylanica*. 135 - 140 In *Pollen: Biology- and Implications for Plant Breeding*. (eds. Mulcahy, D.L. and Ottaviano, E., Elsevier, N.) York 446.

Russell, S.D. (1983). Fertilisation in *Plumbago zeylanica*: Genetic fusion and the fate of the male cytoplasm. *Amer. J. Bot. 70,* 416-434.

Russell, S.D. (in press, quoted by Wilms H.J. and Keijzer C.J. 1985)

van Went, J.L. and Willemse, M.TM. (1984). Fertilisation. pp. 273-317. In : *Embryology of Angiosperms*. (eds. Johri, B.M.) Springer Verlag. Berlin. pp.830.

Wilms, H.J. and Keijzer (1985). Cytology of pollen tube and embryo sac development at possible tools for *in vitro* plant (re)production. pp.24-46 In: *Experimental Manipulation of Ovule Tissues*. (eds. Chapman, G.P., Mantell, S.H. and Daniels, R.W.) Longman, London. pp.272.

4 Dimorphic sperm cell, cytoplasmic transmission and preferential fertilisation in *Plumbago zeylanica*

S. D. Russell

ABSTRACT

Studies of sexual reproduction in the angiosperm, Plumbago zeylanica, a plant lacking synergids, reveals that sperm cells may differ in size, shape, nuclear dimensions and organellar content, and that fertilisation in such a plant may be strongly preferential, reflecting evidence for a final, gamete-level recognition event. One sperm cell of the two in the pollen grain is physically associated with the vegetatitive nucleus and possesses an average of >250 mitochondria and an occasional plastid. The other sperm cell, although connected to the other sperm cell, is not associated with the vegetative nucleus and has 50 mitochondria and an average of 25 plastids. In >94% of the cases examined, the plastid-rich sperm cell preferentially fused with the egg, introducing a highly directed pattern of plastid and mitochondrial transmission.

INTRODUCTION

The occurrence of double fertilisation - an event in which one sperm cell fuses with the egg to form the embryo and the other sperm cell fuses with the central cell to form the nutritive endosperm - is fundamental to sexual reproduction in flowering plants. Genetically, the nuclear complement of the embryo is derived equally from the sperm and egg, but the heritable organelles in its cytoplasm are largely those of the female gamete. The variable contribution made by the cytoplasmic complements of the male and female gamete are strongly influenced by the structure of the gametes at fertilisation and the interaction of these gametes that initiates this process. The initial cytoplasmic constitution of the gametes is crucial in questions of cytoplasmic inheritance. Whether this initial complement is transmitted in entirety, or whether the complement is modified during the course of fertilisation are also important considerations in calculating the actual cytoplasmic contribution of the gametes in the zygote and primary endosperm. Also important are possible differences in the participating cells. This is the underlying rationale for the recent upsurge of interest in the ultrastructure of gametes and fertilisation.

Fertilisation may be divided into three distinct phases: (1) gametic deposition - the release of the male gametes from the delivery structure, the pollen tube; (2) gametic fusion - a cellular event in which the male and female gametes merge to produce a single cell, thus transmitting the nucleus; and (3) nuclear fusion - the mergence of the male and female nuclei. At any of these phases, modifications in the genetic complement of the gametes may occur: in the former two, mechanisms for exclusion of organelles may be present, and in the latter, molecular differences may be present in the male gametes (Day and Ellis, 1984), that may be used to distinguish the male organellar complement by restriction endonuclease mechanisms. Unresolved questions concerning the contribution of the chondriome are: (1) which gametes contribute cytoplasmic organelles and whether these organelles are functional; (2) whether the two gametes delivered to the egg are identical or different in their cytoplasmic constitution; and (3) whether these differences have a preferential impact on patterns of cytoplasmic inheritance.

The possible impact of the paternal chondriome has been one of the last important aspects of fertilisation in angiosperms to be understood, although it was addressed even in the earliest accounts. Strasburger's 1884 description of fertilisation (for review see Maheshwari, 1950) specifically addressed the male cytoplasm and stated that it is not concerned in the process. The earliest work and a common misconception even in modern days suggests that the male gamete lacks a cytoplasm, but by the 1940's evidence for a truly cellular organisation to the sperm cells became overwhelming. Electron microscopic observations have completely confirmed these observations and indicate that the male gamete has a full complement of organelles (for review see Russell and Cass 1981), with one interesting exception: numerous angiosperm male gametes lack plastids (Jensen, 1974; Hagemann, 1976; Kirk and Tilney-Bassett, 1978).

The first genetic evidence that cytoplasmic inheritance did not follow Mendelian patterns in flowering plants was provided by Correns and Baur in 1909 using plastid mutants in four plants. Correns (1909) found that in *Mirabilis, Urtica, and Lunaria* albino plastids were successfully transmitted only by the female parent, but not by the reciprocal mating. Thus, plastid characteristics expected to be borne by the pollen were not transmitted into the female gamete or did not express themselves. Using *Pelargonium*, Baur (1909) proved equally convincingly that the albino plastids present in this plant resulted in a variegated pattern whether the albino parent was the pollen or egg-bearing plant. The female plant, however, did have a slight advantage in the expression of its plastids. These contradictory results reflected two major patterns that predominate flowering plants: a majority of the flowering plants display uniparental inheritance of plastids, as described in the plants crossed by Correns, and a minority of the plants display biparental inheritance, as exemplified in the plants crossed by Baur. The evolutionary placement of the plant appears to have little relationship with whether the plant has a uniparentally inherited plastid complement or whether it is biparental in origin.

Significantly less attention has been given directly to the paternal mitochondrial genome, despite the importance of mitochondria in respiratory metabolism, the synthesis of amino acids, and single carbon metabolism. The majority of cytoplasmic inheritance information in angiosperms has continued to concern plastids because of the crucial importance of chloroplasts to the success of the plant and because of a lack of distinct genetic markers for paternal inheritance of mitochondria (Grun, 1976; Gillham, 1978). Also receiving less attention is the endosperm, which forms as part of a terminal developmental sequence. Although expressing genetic characteristics of the male and female nuclear complement, the participation of a cytoplasmic complement in endosperm has not been thoroughly evaluated. Such formed endosperm is a tissue unique to angiosperms and is an important nutrient source for the developing embryo that may persist up to seed germination and early seedling growth. Both embryo and endosperm are formed by largely similar mechanisms of fusion with the male gamete (Maheshwari, 1950; Jensen, 1974), so participation of the paternal cytoplasmic complement is at least theoretically possible.

As the egg and central cell are large cells they contain numerous

organelles in comparison to those occurring in the sperm cells. The
disproportionate number of maternal cytoplasmic organelles alone may be
a disadvantage to the expression of the male cytoplasmic genome during
embryogenesis. If neither the maternal nor paternal organelles have a
physiological advantage the relative number of organelles may represent
a determining factor in the cytoplasmic constitution of the offspring.
Probability models have been made to explain the composition of
zygotic and derived cytoplasms under differing conditions of relaxed
cellular control (Birky, 1978, 1983). The initial input of such
organelles is of paramount importance in theoretical systems as these
organelles represent the cytoplasmic genetic input of the embryonic
tissues. The actual numerical content of organelles in the gametes
obtained through ultrastructure (Russell and Cass, 1981; Russell, 1983)
can then be useful in predicting the real impact of the paternal cytoplasmic DNA complement in higher plant systems.

Since neither male nor female gametes are presently available in
quantity and both are difficult to manipulate, these cells have largely
been inaccessible to physiological manipulation. In the absence of a
more physiological approach, such information has nonetheless been
obtained through the more time-consuming use of transmission electron
microscopy of fixed cells. Fortunately, it appears that this situation
may soon be changing as a number of laboratories are working actively at
isolating living male and female gametes (Allington, 1985; Zhou and Yang,
1985; Russell, 1986).

This paper presents the clearest view to date of how fertilisation in
flowering plants occurs from results recently obtained using the
angiosperm *Plumbago zeylanica* which lacks synergids. In this plant it has
been possible to describe gametophytic structure in detail and observe the
major phases of gametic interaction which culminate in fertilisation.
In angiosperms with synergids, studies of gametic interaction have been
complicated by the degree of similarity between the cytoplasms of the egg,
central cell, pollen tube, synergids and sperm cells. Further, post-
fertilisation changes in synergids have obscured the fate of the sperm cells
in the past and made light microscopic observations all but impossible
(Maheshwari, 1950). Synergid-less angiosperms, although restricted in
their distribution, make it possible to describe the fate of the male

cytoplasm without the interference of the synergid cytoplasm. Results from synergid-bearing angiosperms are compared in the final section of this paper as a means of integrating these results with the existing literature and providing a vista to some possibilities for the experimental manipulation of fertilisation.

The male gamete

Development

The tissue that eventually forms the male gamete orginates from multiple initials at the reproductive apex and consists of cells derived from both the tunica and corpus of the shoot apex. Up to sixteen micros- porocytes form in each anther sac. Following meiosis, 64 pollen grains can potentially develop in each locule, giving rise to up to 1280 pollen grains per flower. Of these, about 5% will short or appear abnormal, and the number of pollen grains competent to complete pollen tube growth and participate in fertilisation is significantly less than the full number produced. The events of pollen development to the generative cell have been described previously at the light microscopic level (Dahlgren, 1916).

Microspore Division

In *P. zeylanica,* an eccentric mitotic division cleaves the highly vacuolate microspore into a lenticular, densely cytoplasmic generative cell that is appressed to the intine and a large vegetative cell (Russell, 1985a). The generative cell at this stage has the intine as one half of its cell wall. The other half of the cell is delimited by a newly formed wall circumscribing the innermost edge of the cell formed from the cytokinetic phragmoplast. This particular region of the cell wall is initially aniline-blue fluorescent, indicating the presence of callose-like products, similar to the generative cells of other plants that have been described to date (Gorska-Brylass, 1970; Owens and Westmuc- kett, 1983). Elements of the cell wall are also positively stained in Calcofluor white, indicative of $\beta 1$-4 linkages of glucans similar to cellulose. Cellular organelles are not polarized in their distribution at this stage; however, they do appear aggregated. These groupings may originate from naturally occurring increases in the number of organelles arising from replication, possibly representing groups of organelles of common origin.

Soon thereafter, plastids and mitochondria tend to segregate into groups of like organelles in three dimensions. By this time, vacuoles have formed within the generative cell, resulting in a consequent increase in volume and expansion in the thickness of this lenticular cell. The vegetative cell, having essentially reached its mature volume in the microspore stage, displays increases in cytoplasmic volume by displacing space in the large central vacuole. Completion of the replacement of the vacuole does not occur until near maturity of the pollen grain. The vegetative nucleus is large, nearly spherical, conventionally-organized, and contains a prominent nucleolus.

Detachment of the generative cell from the intine occurs by separation of the intine from the edges of the cell, resulting in the generative cell, in essence, peeling away from the edge of the pollen wall (cf. Angold, 1967). This occurs progressively from the edges of the lenticular generative cell to the centre of the cell, until the entire cell is released as a rounded, ellipsoidal cell.

Polarization and morphogenesis

Organelles of the free generative cell (Fig. 1) soon come to be located at opposite ends of the cell, with little overlap in the regions occupied by mitochondria and plastids. The mitochondrial-rich pole, located nearest the vegetative nucleus, elongates into a distinct protuberance that progressively forms a shallow embayment on the surface of the vegetative nucleus (Fig. 1). The protuberance continues to elongate during this developmental stage and concomitantly, the poles of the cell become increasingly distinct. Small vesicles develop in the cytoplasm amid the mitochondrion-rich cytoplasm of the side of the genera-tive cell nearest the vegetative nucleus (Fig. 1). The opposite side of the generative cell contains highly aggregated plastids that by their shear density may deform the intine-facing edge of the generative nucleus (not shown). The generative nucleus, itself, is frequently displaced to the same pole as the plastids.

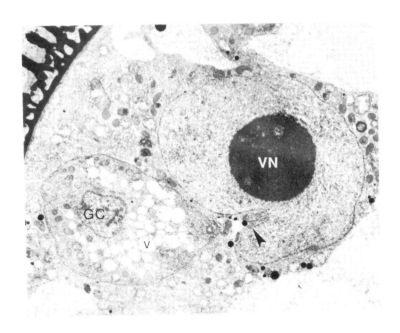

Fig. 1 - Transmission electronmicrograph of the generative cell (GC)
soon after detachment from the intine. The cell soon becomes
polarized with the formation of a protuberance (arrowhead) causing
an indentation in the nearly round profile of the vegetative nucleus
(VN). Vesicles (v) in the generative cell become numerous on the
side of the cell producing the protuberance. X 11,000.

Generative cell division

Elongation of the generative cell reaches its most extreme form just
before the second mitotic division forms the sperm cells. The
protuberance of the generative cell wraps around the vegetative nucleus
prior to division, and essentially reaches its mature size and morphology
prior to division. This projection is nearly 20μm long and tapers from
several micrometers thick near the main sperm cell body to less than one
half of a micrometer at its length. The surface of the projection may be
located within 0.05μm of the nuclear envelope of the vegetative nucleus,
although a direct connection has yet to be demonstrated (Russell and Cass,
1981).

At cytokinesis, the polarized distribution of generative cell organelles is fixed in the maturing sperm cells (Fig. 2). For convenience, these two sperm cells are designated Svn and Sua, for the sperm physically associated with the vegetative nucleus (Svn) and sperm unassociated with the vegetative nucleus (Sua), respectively. As is evident (Fig. 2), the sperm cells are highly polarized, with the plastids located in the sperm cell distant from the vegetative nucleus. Mitochondria predominate in the other cell. Vesicles are numerous in the two cells, disappearing rapidly during later maturation.

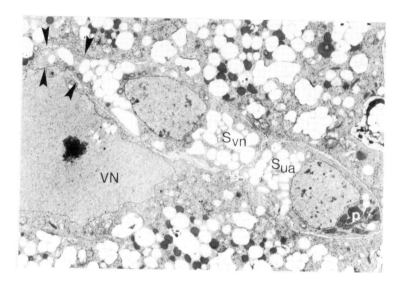

Fig. 2 - Electronmicrograph of the sperm cells soon after generative cell division. One sperm cell (Svn) is physically associated with the vegetative nucleus (VN) by a cellular projection (arrowheads), seen here in oblique section. This sperm cell contains the majority of mitochondria, whereas the smaller sperm cell (Sua) that is unassociated with the vegetative nucleus contains numerous plastids (p). X 4,500.

Mature sperm cell

Cytological organization of the mature sperm cell

The mature sperm cell, as defined in this paper, is present in the pollen grains at anthesis and consists of a full cellular complement including a nucleus, mitochondria, endoplasmic reticulum, ribosomes, Golgi bodies, vesicles, microtubules, and in some sperm cells, plastids (Russell and Cass, 1981; Russell, 1984b). The two sperm cells, differing in size, shape and organelle content, form an example of extreme cytoplasmic dimorphism (Russell, 1984b). These cells may be described as a clear example of cytoplasmic heterospermy, a condition in which the sperm cells clearly display dimorphism in their cytoplasmic genetic complement (Russell, 1985b). A fundamental expression of their polarization is the physical association of the sperm cells with the vegetative nucleus, which is always evident in only one of the two sperm cells.

Quantitative cytology of the mature sperm cell
The Svn is approx. 8 μm long and 4 μm wide at maturity, but tapers into a narrow cellular process up to 30 μm long that wraps around and lies within embayments of the vegetatitive nucleus. This sperm cell contains an average of 256 mitochondria and an average of 0.45 plastids (Table 1.)

Table I   Comparison of sperm cellular organisation in two sperm
           morphotypes of *Plumbago zeylanica* as determined by serial
           sectioning of 11 sperm pairs.   (After Russell, 1984)

Parameter	Svn[a]	Sua[a]
Mitochondria	256.2	39.8
Plastids	0.45	24.3
Microbodies	0.36	3.2
Cell volume[b]	69.5	48.9
Cell surface[c]	147.9	84.7
Nuclear volume[b]	19.9	12.1
Cytoplasmic volume[b]	45.2	33.4

Abbreviations: Svn = sperm physically associated with vegetative
                nucleus; Sua = sperm unassociated with vegetative
                nucleus
                [b] measured in cubic micrometers
                [c] measured in square micrometers

Ultrastructural modifications to organelles

This is the larger cell, having an average volume of 69.5 μm^3, an average surface area of 147.9 μm^2, and an average nuclear volume of 19.9μm^3 or 28.6% of the cellular volume (Russell, 1984).

The sperm cell unassociated with the vegetative nucleus (Sua) is also approximately 8μm long and 4μm wide, but lacks an extended protuberance. It contains an average of 24 plastids and 39.8 mitochondria (Table 1). The Sua is the smaller of the two sperm cells, having an average volume of 48.9μm^3, an average surface area of 84.7μm^2, and an average nuclear volume of 12.1μm^3 or 24.7% of the cellular volume (Russell, 1984b). Sperm mitochondria have an average minimum diameter of 0.286μm and are spheroidal to ellipsoidal (Russell, 1983).

Three-dimensionally, the two sperm cells are located with their axes at approximately 60° to one another, enclosed together within the inner vegetative cell plasma membrane, and joined by a common cell wall traversed by plasmodesmata. Although plasmodesmata are evident between the two sperm cells, and at earlier stages, between the sperm and vegetative cell, the cell wall diminishes to the point that the presence of plasmodesmata in the sperm-vegetative cell walls is difficult to demonstrate (Russell and Cass, 1981). Perhaps such connections are severed at anthesis and the reproductive cells remain separated from the cytoplasm of the growing pollen tube. The polarity of organelles observed at generative cell division is fixed within the sperm cells at maturity (Fig. 2). Although mitochondria appear randomly distributed within the sperm cells, they are in reality aggregated in three dimensions, as apparently are some of the plastids (unpublished data). When plastids infrequently occur in the Svn they occur near the crosswall between the two sperm cells, providing evidence for a continued polarization of cytoplasmic components in the sperm following generative cell cytokinesis (Russell, 1984b). Mitochondria appear to occur as frequently in the main body of the Svn as they do in the cellular projection (Russell, 1984b).

Comparison of the sperm cell with the vegetative cell

In comparison with the volume of the vegetative cell, which is on the order of 110,000μm^3, with a surface area of at least 5,9000μm^2, the sperm cells represent a small potential reservoir of organelles, each a mere 1/1900 of the volume of the vegetative cell. The cytoplasm of the pollen grain, however, becomes so greatly modified upon reaching anthesis, that it has a preponderance of vesicles, less numerous mitochondria and relatively few plastids. The mitochondria are an average of 0.453μm in minimum diameter and are spherical to roundly ellipsoidal. The plastids, previously swollen with starch, have an average sectional area of $0.785~\mu$m^2 at anthesis. These characteristics are all indicative of a remarkable specialization of the vegetative cell cytoplasm that is mobilized with the occurrence of germination on the stigma (cf. Jensen and Fisher, 1970).

The mass of an individual pollen grain of *P. zeylanica* at maturity is on the average 137 ng, with the sperm cells representing a mean 72 pg of this mass. Despite the relatively small size of these gametes, presumably these cells have become modified in such a means as to favour their successful arrival at the female gametophyte and specifically fuse with the female reproductive cells. Among these modifications for successful transmission may be the occurrence of a physical association between the sperm cells and one between a sperm cell and the vegetative nucleus, forming a functional entity termed the male germ unit (Dumas, Knox, McConchie and Russell, 1984). Under the rapid conditions of pollen tube growth, the association of sperm cells in register would promote their rapid co-transmission into the female gametophyte at a time when as a result of pollen tube discharge, conditions within the female cells could be rapidly changing. Additionally, this would practically assure against heterofertilization, a condition in which sperm cells originating from different pollen tubes fertilize the egg and central cell.

The unique properties of sperm cells have yet to be fully understood and exploited, but may provide an additional haploid source for androgenic experiments and gametophytic screening. The isolation of sperm cells is technically challenging, but has been done using *Plumbago* (Russell, 1986) and has been proposed for other plants. Characterization of the unique properties of these cells will require the ability to isolate them in the living state, purify them, culture them, and eventually regenerate them into green plants. *Plumbago* may be a particularly interesting plant in this regard because the two dimorphic sperm cells, as a result of their vast differences in organelle content, may regenerate highly dissimilar plants.

FEMALE GAMETE STRUCTURE

Cytological organization

The female gametophyte of *Plumbago* (Fig. 3) is organized into five cells: the egg, two lateral cells, an antipodal cell and a proendospermatic central cell with its nucleus formed by the fusion of four polar nuclei, thus conforming to the *Plumbago* type of megagametophyte development (Maheshwari, 1950). This type of megagametophyte organization is seen only in the subfamily *Plumbaginoideae* (family *Plumbaginaceae*) consisting of the genera *Ceratostigma, Dyerophytum* (formerly, *Vogela*) and *Plumbago*.

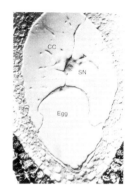

Fig. 3 - Differential intereference contrast micrograph of the female gametophyte of *Plumbago zeylanica*, illustrating the highly vacuolate egg and central cell (CC). The secondary nucleus (SN) of the central cell is the product of the fusion of four polar nuclei. X 150.

An even more highly reduced form of megagametophyte development occurs in *Plumbagella*, in which the lateral cells are never formed. In this monotypic genus, a secondary fusion of the three chalazally-placed megaspores results in a triploid nucleus which undergoes a single subsequent mitotic cycle to produce a triploid antipodal cell and a triploid polar nucleus which contributes to the formation of a tetra-ploid central cell. The mature megagametophyte therefore consists of only an egg, central cell and one antipodal (Maheshwari, 1950). Interestingly, this is the closest approach to an animal-like form of reproduction that has been reported in flowering plants and approaches the level of reduction that occurs in *Fucus,* a brown alga, which has the most highly reduced reproductive structures of the parenchymatous plants. Fertilization events in *Plumbagella* are essentially identical to those in *P. zeylanica* (Cass and Russell, unpublished data).

The egg of *P. zeylanica* is a prominent and highly vacuolated cell, with a chalazally-placed nucleus located adjacent to the nucleus of the central cell (Fig. 3). The egg is enclosed with a continuous cell wall (Cass and Karas, 1974; Russell, 1982), with the upper walls occupied by a largely non-fibrillar wall material, which is readily displaced by the discharge of the pollen tube prior to fertilization (Russell, 1982). The micropylar end of the egg is occupied by a filiform apparatus – a collection of cell wall ingrowths characteristic of the synergid in conventionally-organized female gametophytes. The presence of a filiform apparatus with aggregated mitochondria and microbodies suggests a greater level of physiological activity in the eggs of *Plumbago* than occurs in more typical angiosperm eggs (Cass and Karas 1974). Lacking synergids, the egg of *Plumbago* has undertaken a number of typical synergid functions including that of receipt of the pollen tube (Russell 1980, 1982). As the egg receives the pollen tube more directly and may participate more extensively in its own nutrition than the conventionally-organized angiosperm egg, there is strong evidence for a direct transference of synergid function to the egg.

The lateral cells, which are the only developmental counterpart to synergids in their mode of origin, located high on the megagametophyte walls, are spatially removed from the process of fertilization (Russell, 1981, 1982). They may degenerate prior to fertilization, or remain functional into early embryogenesis. These and the antipodal cell may in some anomalous megagametophytes become greatly hypertrophied (see Dahlgre, 1916, 1937), but this structural oddity does not appear to interfere with fertilization. Apparently, such cells are relatively more long-lived than their smaller counterparts, but still have no more distinct function than their smaller, more typical accessory cells. Although these cells may reach the same size as the egg, they may still be distinguished by their placement in the megagametophyte and their consistent absence of a filiform apparatus.

Ultrastructural modifications to organelles

The central cell contains a prominent secondary nucleus which is the fusion product of four polar nuclei (Fig. 3). Central cell cytoplasm is located peripherally at the edges of the megagametophyte, with numerous transvacuolar strands (Fig. 3) traversing the prominent vacuole and a small region of perinuclear cytoplasm surrounding the secondary nucleus and extending to the egg. Plastids tend to remain in a perinuclear position in the mature megagametophyte prior to fertilization, although mitochondria are distributed more evenly throughout the cell.

Table II   Comparison of cellular organisation of egg and central
           cell (proedosperm) of *Plumbago zeylanica* (unpublished data)

Parameter	Egg	Central Cell
Mitochondria	14,340	75,500
Plastids	110	300
Cell volume[a]	440,000	2,400,000
Cell surface[b]	26,200	85,100
Nuclear volume[a]	940	3,400
Cytoplasmic volume[a]	31,600	299,000

[a] measured in cubic micrometers
[b] measured in square micrometers

Quantitative cytology

The female gamete and central cells (Table 2) are far more voluminous cells than the male gametes and pollen tube (cf. Table 1). The vast majority of the volume of the female cells is occupied by large vacuoles that represent over 90% of the volume of the egg and nearly 85% of the volume of the central cell. The nucleus of the central cell is approximately four times larger than that of the egg and is located in a cell that is over five times larger than the egg.

The quantity of organelles present was estimated by stereology (Weibel, 1979). The volume fractions occupied by each organelle type were determined in representative regions of the megagametophyte and then divided by the volume of an average organelle as determined from serial

ultrathin sections. The number of mitochondria and plastids in the egg and
central cell were far in excess of the normal amount found in somatic cells.
In a typical somatic cell, the number of mitochondria would be expected
to be significantly less than 500 and the number of plastids less than
75. A preliminary count of heritable organelles in the egg and central cell
indicates that there are over 14,000 mitochondria and 110 plastids in the
egg and over 75,000 mitochondria and 250 plastids in the central cell
(Table II)

FERTILIZATION AND ORGANELLE TRANSMISSION

Sperm deposition

Following eight and one-half hours of pollen tube growth after pollination,
the tip of the pollen tube pushes out of the stylar transmitting tissue,
penetrates between the integuments and enters the micropyle. Within the
nucellus, enzymatic separation of the middle lamella permits the tube
with the male gametes within it, to penetrate into the female gametophyte,
and to finally push between the egg and central cell to deposit the two
sperm cells (Russell 1981, 1982). When the pollen tube elongates up to
70-80μm within the megagametophyte, a aperture forms at the tip of the
pollen tube (Fig. 4a). With the weakening of this cell wall, the plasma
membrane of the pollen breaks (Fig. 4b), two single membrane-bound sperm
cells are released (Fig. 4c), and the contents of the pollen tube, including
the sperm and vegetative nucleus are released between the egg and the
central cell (Fig. 5a). The outer membrane (inner pollen plasma
membrane), present during all stages in development prior to gametic
deposition, is removed during discharge of the pollen tube (Russell 1983),
and may represent for the sperm cells, a phase analogous to capacitation
in animal systems. Membranes of the sperm cells contact both the egg and
central cell at this phase (Fig. 5a). When the Svn is expelled from the
pollen tube, the cellular protuberance that physically associates it with
the vegetative nucleus is severed and its torn segments spontaneously form
small enucleated cytoplasmic bodies (Figs. 4b and 4c), which do not
participate in fusion (Russell, 1983). These bodies may contain as much
as 15% of the cytoplasmic organelles of the Svn, and constitute one
method by which organelles may be excluded at the time of gametic deposition

(Russell, 1983). The gametic deposition phase is accomplished quickly, as evidenced by rare instances that have been reported in the literature (see Russell, 1983). Probably this brief phase represents less than 30 seconds of the more than eight hours after pollination required to effect fertilization in *P. zeylanica*.

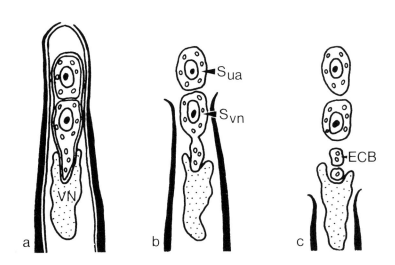

Fig. 4 - Reconstruction of events in *Plumbago zeylanica* occurring at the pollen tube during gametic deposition. (a) Sperm cells and vegetative nucleus (VN) move toward the weakening apex of the pollen tube. (b) Sperm cells (Sua and Svn) are released from the outer delimiting pollen plasma membrane and separate from one another. (c) As sperm cells leave the pollen tube, the cellular projection that physically associates the Svn with the vegetative nucleus is severed, forming enucleated cytoplasmic bodies (ECB), identifiable because of the presence of characteristic sperm mitochondria. In actuality, either sperm cell may exit first and only the Sua exiting first is illustrated here. (After Russell, 1984a).

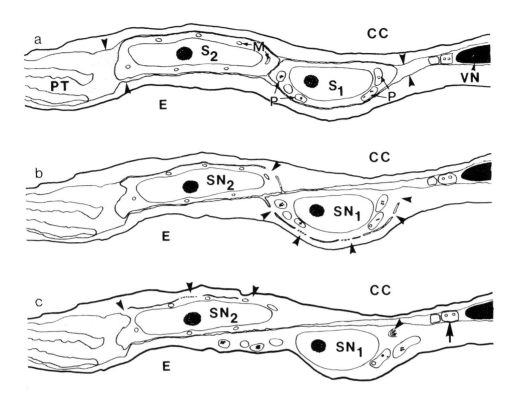

Fig. 5 - Reconstruction of gametic fusion in *Plumbago zeylanica*.
(a) Sperm cells (S1 and S2) are deposited by the pollen tube
(PT) within the female gametophyte between the egg (E) and central
cell (CC) as single membrane-bound cells with numerous heritable
cytoplasmic organelles (M, mitochondria; P, plastid). Vegetative
nucleus (VN) segregates from the sperm cells and then degenerates
between the egg and central cell. (b) Fusion is initiated first at
a single site, as indicated by the arrowhead in the sperm cell
occupied by SN2. Subsequently, as shown by arrowheads near the
other sperm nucleus (SN1), additional fusion sites form rapidly,
followed by progressive vesiculation of the remnant sperm membranes
and transmission of the contained organelles of the sperm cell into
the female reproductive cells. (c) The completion of gametic fusion
is marked by the disintegration of the fusion membranes into a
transient whorl (seen near arrowhead in egg). Enucleated cytoplasmic
body originating from the Svn is indicated by unlabelled arrow.
(After Russell, 1983.)

Ultrastructural modifications to organelles

Gametic fusion

Gametic fusion is initiated by a single fusion event between membranes
of the sperm cells (Fig. 5b) and their respective female reproductive cells.
Subsequently, a number of secondary fusion points between the two cells
occur independently at the surface (Fig. 5b) and this is concluded by
complete disorganization of the membranes (Fig. 5c). Gametic fusion
results in the transmission of male gamete nuclei with co-transmission of
nearly the entire sperm cytoplasmic complement. It is possible that this
entire process is completed within 15 to 40 seconds and the primary
fusion event is so transient that it is probably accomplished within one
second, similar to other biological systems studied to date (daSilva and
Kachar, 1980). Possibly electron microscopists will never be able to
capture the primary fusion event of flowering plant reproduction in vivo
despite the degree of interest in visualizing this event.

Mitochondrion and plastid transmission

Further evidence for the plasmatic nature of gametic fusion is provided
by the presence of ultrastructurally identifiable paternal organelles
within the newly fertilized zygote and primary endosperm. Paternal
mitochondria are smaller in size than those originating in the maternal
tissue and can be demonstrated on a populational basis by statistical
methods. Mitochondrial width serves as the most easily compared parameter
of size as it is least sensitive to sectioning orientation. Although the
shape of both maternal and paternal mitochondria is spheroidal to
ellipsoidal, mitochondrial size differs with respect to cellular physiology
and can be statistically discriminated in different cell types. Mitochondrial size may be used to identify the origin of mitochondria, including
those of enucleated cytoplasmic bodies and in female reproductive cells
soon after their transmission.

Mitochondrial widths of eight different cellular origins are
characterized by differing widths (Table III) that are statistically
different on a populational basis (Table IV). Mitochondria originating in
maternal tissue are significantly larger than those originating in the
paternal tissue of sperm cells and the growing pollen tube. Egg and
central cell mitochondria are not significantly different in width (Table IV),

although sperm mitochondria and those within the growing pollen tube are significantly smaller (Tables III and IV) and can be identified adjacent to the sperm nucleus after fusion. Such mitochondria, located within 1μm of the sperm nucleus (labelled 'CC.SN' in Tables III and IV), are typified by mitochondrial widths that differ insignificantly with those originating from the sperm in the actively growing pollen tube (Tables III and IV). Small enucleated cytoplasmic bodies, presumably originating from the breakdown of the sperm projection associating the Svn with the vegetative nucleus, have mitochondrial widths that are statistically comparable with sperm-originating mitochondria. Very significant differences are noted in comparing mitochondrial width in the enucleated cytoplasmic bodies and transmitted paternal organelles with that of mitochondria in the pollen tube and female reproductive cells (Table IV).

Table III   Mitochondrial widths of selected reproductive cells of
            Plumbago zeylanica in micrometers.   (After Russell, 1983.)

Cellular sources[a]	Mean	SD	0.95 Confidence interval
Pollen grain (anthesis)	0.453	0.040	(0.437 - 0.469)
Pollen tube (near ovule)	0.433	0.062	(0.416 - 0.451)
Sperm (in pollen grain)	0.286	0.042	(0.271 - 0.302)
Sperm (in pollen tube)	0.224	0.034	(0.206 - 0.242)
Egg	0.623	0.137	(0.592 - 0.654)
Central cell	0.489	0.088	(0.460 - 0.519)
Fertilized CC near SN	0.214	0.037	(0.203 - 0.226)
Enucleate cytoplasmic body	0.203	0.061	(0.155 - 0.250)

[a] Pollen tube, sperm in growing pollen tube, egg and central cell mitochondrial widths taken from gynoecia in which pollen tubes had penetrated to within 100μm of the egg.   CC = central cell, SN = sperm nucleus.

Table IV   Comparison of mitochondrial widths between mitochondria of
eight different cellular origins in *Plumbago zeylanica*.
Statistical differences are result of paired three-nested
ANOVA calculations, except as noted[a] (After Russell, 1983.)

	PG	PT	E	CC	S.PG	S.PT	CC.SN	ECB
PG		ns	ns	ns	**	**	**	**
PT	ns		**b	ns	**	**b	**	**
E	ns	**b		**b	**	**	**	**
CC	ns	ns	**b		**	ns	ns	ns
S.PT	**	**b	**	**	ns		ns	ns
CC.SN	**	**	**	**	ns	ns		ns
ECB	**	**	**	**	ns	ns	ns	

[a] Abbreviations used in Table II: CC = central cell; CC.SN = central cell near sperm nucleus; E = egg; ECB = enucleate cytoplasmic body; PG = pollen grain; PT = pollen tube; S.PG = sperm in pollen grain at anthesis; S.PT = sperm in pollen tube <100 m from ovule. Statistical abbreviations: * = P<0.05; ** = P<0.01; ns = not significant.

[b] Paired data from same electron micrographs.

The transmission of sperm plastids can be demonstrated directly through ultrastructure because of numerous differences in the structure of paternal and maternally-originating plastids (Table V); (Russell, 1980, 1983, 1985b). Paternal plastids in the fertilized egg may be distinguished by smaller size, more pleiomorphic shape, greater stromatal density, less constricted lamellar membranes and more random distribution of plasto-globuli (Table V). Both maternal and paternal plastid populations are visible in the egg soon after gametic fusion (Fig. 6); (Russell, 1980, 1983). These differences are striking enough that the organelles may be readily identified in the early stages of embryogenesis. Apparently, all sperm plastids present in the sperm cells are transmitted into the female reproductive cells during fusion as the identification of these plastids is possible following fertilization and none have been identified outside of the egg and central cell.

Table V   Differences in organellar morphology between plastids of sperm
          origin and those of the megagametophyte.   (After Russell, 1985)

Characteristic	Sperm plastid	Egg plastid
Size	up to 1.5μm	up to 2.5μm
Ave. sectional area	0.518μm^2	0.893μm^2
Shape	Ellipsoidal, elongate	Pleiomorphic
Stroma	Denser than cytoplasm	Equally electron dense
Lamellae	Inflated	Constricted
Plastoglobuli	Aggregated	Randomly distributed

   The cytoplasmic contribution of the female reproductive cells (Table II)
makes any contribution of mitochondria by the sperm cells (Table I) seem
insignificant by comparison, except in the unlikely event that paternal
organelles are for some reason favoured in organellar competition or
genetic recombination.   There are *ca* 14,000 mitochondria in the egg and
ca 75,000 mitochondria in the central cell (Table II) compared to 256 and
40 mitochondria in the two sperm morphotypes.   Similarly, the 300 plastids
of the central cell seem to dwarf the possible contribution of sperm
plastids.   The contribution of male plastids in the egg, although decidedly
biased by the more numerous maternal plastids (Table II) and the larger
individual size of maternal plastids (Table V), might nevertheless gain
expression in the mature plant.   The ratio of egg plastids to sperm
plastids of 2:1 to 4:1 (female:male) is approximately the same as that
occurring in *Oenothera erythrosepala* (Meyer and Stubbe, 1974), a plant in
which plastids are known to be biparentally inherited.   Interestingly, the
paternal plastids of *Oenothera* are smaller in volume than maternal plastids
(Meyer and Stubbe, 1974), as has been reported in *Plumbago* (Russell, 1983),
so it appears that size differences alone may not limit the competitiveness
of paternal plastids.

   Nuclear fusion
Although sperm nuclei are initially deposited 20-30μm from their respective
female nucleus, within minutes they become aligned with the appropriate

female nucleus, whether it be the egg or proendospermatic nucleus, and become appressed to it. These nuclei appear to travel the shortest possible
distance to the female nuclei and may reach the female nuclei within
several minutes after gametic fusion. In *Plumbago*, fusion occurs rapidly
 enough that intermediate stages in fusion are rarely seen (Fig. 6);
(Russell, 1983), but in cotton, it appears that fusion is initiated in
the outer nuclear envelope at several points almost simultaneously, which
is then soon followed by fusions between membranes in the inner layer of
the nuclear envelope (Jensen 1974). Presumably, this is also the sequence
of events in *Plumbago*. Subsequently, "nuclear bridges" are formed between
the male and female nuclei linking the nucleoplasm of the two cells
(Jensen, 1974; Russell, 1983).

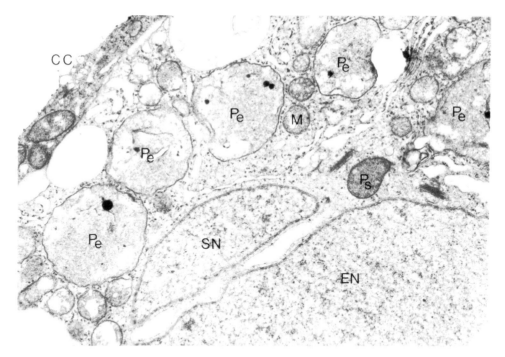

Fig. 6 - Transmission electronmicrograph of a fertilized egg of *Plumbago*
         zeylanica, showing the presence of both maternally-originating
         plastids (Pe) and paternally-originating plastid (Ps) near the
         sperm nucleus (SN) and egg nucleus (EN). These nuclei have
         already fused as visible in other sections. CC = central cell;
         M = mitochondrion. X 13,500.

Fate of pollen tube cytoplasm

Gametic fusion results in the transmission of the sperm nuclei and their surrounding sperm cytoplasm into the female reproductive cells (Russell 1980, 1983), but apparently excludes the rapidly degenerating pollen tube cytoplasm. The pollen tube organelles probably do not participate in cytoplasmic transmission, judging by their degenerate ultrastructural appearance after discharge from the pollen tube. As the pollen tube bursts, its delimiting membrane is ruptured and its organelles are no longer in a cellular environment that is protected from the extra-cellular hydrolysis and the abundant aldehydes known to be present in cell walls and vesicles within the angiosperm pollen tube before discharge (Russell, 1982; Fisher and Jensen, 1969). Mitochondria and plastids released from the pollen tube are morphologically altered within seconds of their release as evidenced by both freeze-substitution (Russell, 1982) and chemical fixation (Russell, 1982, 1983). As the extracellular matrix may also contain various nucleases, the tube organelles are presumably inactivated in the intercellular space between the egg and central cell while fertilization is occurring within the egg. The intact egg and central cell plasma membranes are capable of excluding the discharged organellar contents of the pollen tubes.

Preferential fertilization

Since the sperm differ significantly in organelle content in *Plumbago*, it is possible to distinguish which sperm cell fused with the egg or central cell by identifying sperm plastids in the female reproductive cells. Since plastids can be directly identified to determine their cellular origin, these provide an internal tracer to the fate of the plastid rich sperm cell (Sua). Three conditions must be met, however, for this to be a useful criterion for determining cell fate. First, the patterns of plastid distribution in the dimorphic sperm cells must be consistent in each morphotype, which has been clearly shown in *Plumbago* (Russell, 1984b). Second, one consistent plastid population must be present in the egg prior to fertilization (shown in Cass and Karas, 1974; Russell, 1982) and both populations of plastids must be present in the fertilized egg (Russell, 1980, 1983, 1985b). Thirdly, the degree of natural variability in the number of plastids occurring in the plastid-poor sperm cell (Sua) must be determined to predict the likelihood of error and to set a numerical criterion for accepting a dataset. The likelihood of plastids being transmitted into a female cell by the Svn is given by a Poisson distribution (Table VI). If a 1% possibility of misidentification is acceptable, then finding three sperm plastids in a given fertilized egg will meet the criteria for identifying which of the sperm cells fertilized the egg.

Table VI Frequency of plastids present in the Svn, based on a Poisson distribution of absolute organelle counts of eleven sperm cells. (After Russell, 1984, 1985.)

No. Plastids	Frequency
0	0.6347
1	0.2885
2	0.0658
3	9.9351×10^{-3}
4	1.1290×10^{-3}
5 or more	1.1095×10^{-4}

After considerable work, 17 datasets were accumulated on this question and in 16 of these, the plastid-rich Sua was proven to have fertilized the egg, whereas the mitochondrion-rich Svn fused with the central cell in only one case. The probability of this ratio occurring by chance alone, according to the binomial distribution is less than 1 in 7000. Therefore, random fertilization is highly unlikely in *Plumbago* and preferential fertilization is far more likely to have occurred (Russell, 1985b). Since the opposite pattern of fusion occurred in one case, this pattern is best described as being a preferential rather than an entirely specific event occurring in about 94% of the cases studied (Russell, 1985b). As abortion of approximately 5% of the ovaries is typical, it is possible that the atypical pattern of fusion may represent one source of ovule abortion occurring in outwardly normal flowers.

Based on this data it seems reasonable to propose that a recognition system is in effect in the megagametophyte of *P. zeylanica*. Presumably, similar to recognition phenomena in other biological systems (Clarke and Knox, 1978), it functions at the plasma membrane level and may include complementary elements of recognition in the surface of both the egg and sperm cells. Contrary to most selective fusion mechanisms in other biological systems, however, distinction between the two morphotypes in *Plumbago* may occur though means of a threshold effect whereby the highly significant differences in sperm surface area (Table I); (Russell, 1984b) may provide the basis for selective fusion.

Where the specificity of sperm cells arises by a single recognition factor or by multiple recognition factors is crucial to understanding fertilization in this angiosperm, but is a difficult question to determine by conventional embryological methods. Reciprocal receptors for different sperm morphotypes cannot be easily reconciled with the fusion of the plastid-rich sperm with the central cell in the one reported exception to the general pattern in *Plumbago*. If progress is to be made on this problem, it will be necessary to characterize the surfaces of both the egg and sperm cells to determine their cellular specificities.

Concluding remarks on *Plumbago* fertilization

The degree of complementary detail available on the gametes and sexual reproduction in *Plumbago zeylanica* is presently the greatest available today in any flowering plant. It suggests that gametic fusion, far from being a random event, is one in which fusion events are stringently programmed and cytoplasmic organelles are transmitted in a conservative manner. Sperm cells are deposited within the female gametophyte with an accuracy of less than 10μm inthe location in which they are deposited. The pollen plasma membrane surrounding the sperm cells as they travel in tandem, physically associated with the vegetative nucleus is retained throughout pollen tube growth up to the time of gametic discharge, at which time the actual sperm surface is first uncovered to the environment at the surface of the egg and central cell simultaneously. Further, their fusion is preferential ànd not random in any sense. All of the Sua is transmitted; most, but not all of the Svn is transmitted.

Although further exploitation of this system is clearly necessary in order to understand the subcellular and molecular basis of double fertilization, the level of detail already reported for this plant is needed for all of our crop angiosperms if man is to benefit fully from the potential of genetic engineering. Available in the naturally-occurring mechanism of double fertilization in angiosperms is the possibility of introducing a genetic element with such subtlety that the egg or central cell may be separately altered in its characteristics. Presumably, any such mechanism will be based on differences in sperm cells, which are only beginning to be explored. The development of an *ex ovulo* method of *in vitro* fertilization may be desirable in exploiting these differences. Further techniques could be developed to provide either nuclear, cytoplasmic, or plasmid-based heterospermic male gametes within essentially normal pollen capable of delivering the sperm cells through normal reproduction. If this were to be accomplished, one could envision this as the use of the ultimately most specific genetic vector to crop plants that could be applied easily and safely to plants grown in agricultural settings. The extent to which the findings on fertilization, and specifically gametic fusion, in *Plumbago* may be applied to other angiosperms is limited at present because of a lack of detailed information in synergid-containing species.

SIGNIFICANCE OF THE *PLUMBAGO* STUDY WITH THOSE CARRIED OUT ON OTHER ANGIOSPERM SPECIES

Generative cell

The most thoroughly studied aspects of generative cell structure have included the nature of the cell wall and the presence or absence of plastids. Few studies have considered the entire course of development of the generative cell, but among them the most complete are studies of *Beta* (Hoefert, 1968, 1969, 1971), *Endymion* (Angold, 1967; Burgess, 1970), *Haemanthus* (Sanger and Jackson, 1971a, 1971b, 1971c), *Hordeum* (Cass and Karas, 1975) and massulate orchids (Heslop-Harrison, 1968).

Generative cell cytology

The generative cell is normal cell, in a cytological sense, except that plastids are commonly absent. Two mechanisms have been proposed for their elimination. The first is simple "exclusion" (Hagemann, 1976), in which plastids are physically excluded during post-meiotic mitosis. The second possibility is that the plastids are selectively eliminated later in generative cell development (Clauhs and Grun, 1977; Hagemann, 1979). The latter elimination could be mediated by molecular modification of the paternal plastid genome (Day and Ellis, 1984) followed by selective lysis during generative cell development. However, the presence or absence of paternal organelles in the generative cell appears to correlate closely enough with plastid inheritance in the embryo, that post-generative cell mechanisms for the elimination of plastids (Vaughn, *et al.*, 1980) may be unimportant under normal circumstances. The majority of flowering plant generative cells examined to date lack plastids (Kirk and Tilney-Bassett, 1978) and correspondingly, have uniparental maternal inheritance of plastids. In a minority of angiosperms, generative cells contain plastids and biparental plastid inheritance occurs (Kirk and Tilney-Bassett, 1974).

Generative cell morphogenesis is initiated by the presence of numerous longitudinally-oriented microtubules localized near the surface of the generative cell. These are present in the elongating projection and extend into the cortical region of the main cell body. Such microtubules have been implicated in the morphogenetic processes occurring during generative cell maturation (Sanger and Jackson, 1971b), the formation of sperm cells (Cass, 1973; Hoefert, 1969), and may also have a role in development during later maturation, including passage in the pollen tube (Cass, 1973; Cresti, Ciampolini and Kapil, 1984). Perhaps the best demonstration of this phenomenon is that of Sanger and Jackson, where the addition of colchicine altered the form of immature generative cells and the spindle-shaped mature stage never formed in the presence of microtubule-altering drugs. Microtubules present in the immature generative cell have also been described as participating in the formation of a pre-prophase band (Owens and Westmuckett, 1983); however, the cortical microtubules described therein appear to be oriented at right angles to the plane of the future phragmoplast.

Generative cell wall

Perhaps the most controversial topic in generative cell structure has been the presence or absence of a generative cell wall and its possible structure. Among those considering that no cell wall is present include Bopp-Hassenkamp (1960), Larsen (1963, 1965), and in the later sperm cell there is general support for the absence of a cell wall (Cass, 1973; Cass and

Karas, 1975; Hoefert, 1969, 1971; Jensen and Fisher, 1968b; Karas and
Cass, 1976) or at least an unconventional cell wall composition excluding
fibrils (Russell and Cass, 1981). The majority of the studies support
the presence of a modified cell wall surrounding the generative cell
(Angold, 1967; Burgess, 1970; Cocucci and Jensen, 1969; Dexhelmer, 1965;
Diers, 1963; Dupuis, 1972; Gorska-Brylass, 1970; Heslop-Harrison, 1968;
Horvat, 1969; Jensen and Fisher, 1970; Lutz and Sjolund, 1973; Maruyama,
Gay and Kaufman, 1965; Owens and Westmuckett, 1983; Roland, 1971; Sassen,
1964; Sassen and Kroh, 1974).

The work of Gorska-Brylass (1970) and Owens and Westmuckett (1983),
in particular, describe the generative cell wall as a dynamic structure
undergoing transient chemical changes. Initially, the cell wall is formed
through a conventional cell plate (Karas and Cass, 1976), but almost
contemporaneously with its inception, an aniline blue-fluorescent component
(presumably callose) is incorporated into its structure (Gorska-Brylass,
1970; Owens and Westmuckett, 1983). This is a transient component which
reportedly diminishes during detachment of the generative cell from the
intine (Gorska-Brylass, 1970; Owens and Westmuckett, 1983). The latter
study also notes that the wall temporarily binds Calcofluor white M2R as
well, indicating the possible presence of $\beta 1$, 4-linked polysaccharides.
The generative cell wall continues to react positively using the PAS
reaction (see below) even at the time of pollen dispersal from the flower,
although it apparently loses the ability to bind Calcofluor white.

Burgess (1970) reported that although a wall may appear to be present
around the generative cell, it has "no inherent rigidity" and may not
contain noticeable fibrils. This has since been confirmed by other
researchers (Sassen and Kroh, 1974; Russell and Cass, 1981). Russell and
Cass (1981) regard the boundary layer of the sperm cells of *Plumbago*
as a polysaccharide-rich layer, which can be cytochemically visualized
in freeze-substituted material, but is lacking in the fibrillar components
normally present in conventional cell walls. The wall layers appear to be
poorly cross-linked since no insoluble polysaccharides could be identified
in the mature sperm cell after conventional chemical fixation using either
the PAS method or the PA-TCH-SP reaction of Thiery (1967). Russell and
Cass (1981) further noted that the amount of cell wall material present
in freeze-substituted material diminishes during sperm cell passage in

the pollen tube, and that the amount of cell wall present at the time of gametic fusion is negligible (Russell, 1983).

Sperm cell cytology

In ultrastructure, the sperm cells of angiosperms appear to be conventionally organized cells containing a nucleus, plasma membrane, mitochondria, ribosomes, endoplasmic resticulum, Golgi bodies, microtubules and the typical density of ground cytoplasm observed in somatic cells (Cass, 1973; Cass and Karas, 1975; Dumas, *et al.*, 1985; Hoefert, 1969; Jensen and Fisher, 1968b; McConchie, Jobson, and Knox 1985; Russell and Cass, 1981; Zhu, Hu, Xu, Liandsen, 1980). Microfilaments have been reported in sperm cells of two species (Russell and Cass, 1981; Zhu, *et al.*, 1980) and microbodies in one (Russell and Cass, 1981; Russell, 1984b). Mitochondria are reported to be absent in one species of orchid (Coccuci and Jensen, 1969), although it is possible that such organelles are either so highly dedifferentiated as to render them unidentifiable (vesicles of approximately the same size as mitochondria are present in the cytoplasm of this species) or the numbers of mitochondria are so severely reduced that encountering one is a matter of chance.

Sperm cells are structurally distinct from other somatic cells in three important regards. First, the sperm cells lack a highly organized cell wall. Although the presence of polysaccharides can be demonstrated in the mature sperm cells by the use of freeze-substitution (Russell and Cass, 1981), microfibrils are not present in these layers. The surface is ultrastructurally more similar to a glycocalyx (Russell and Cass, 1981; Dumas, *et al.*, 1985) as is observed in animals or a periplasmic space (McConchie, *et al.*, 1985). Throughout later pollen tube development, the sperm cell surface appears to decrease in thickness prior to fertilization. When the sperm cells are released from the confines of the pollen cytoplasm, sperm cells are altered in morphology, becoming more rounded and exhibiting slow shape changes (Cass, 1973; Russell and Cass, 1981) from round to slightly ellipsoidal (Cass, 1973). These isolated cells clearly demonstrate that sperm cells are, in essence, wall-less cells and one of few examples of naturally occurring plant protoplasts (Dumas, *et al.*, 1984). Second, sperm cells appear to be surrounded by two plasma membranes rather than one. The inner membrane is the sperm plasma membrane, while the outer is an inner plasma membrane of the pollen grain vegetative cell (Jensen, 1974). In fact, the sperm would outwardly appear to be a conventionally-organized cells were a cell wall present. That sperm cells occur totally enclosed within the vegetative cell is of inherent biological interest as few examples of this kind of behaviour can be given in biology. Third, plastids may or may not be present in the mature sperm cells. A previous survey of the literature revealed that biparental inheritance of plastids occurs in less than 1/3 of angiosperm species examined (Kirk and Tilney-Bassett, 1978). In over 2/3, the offspring displayed uniparental maternal inheritance of plastids (Kirk and Tilney-Bassett, 1978). Observed patterns of plastid inheritance appear to relate ultrastructurally to the presence or absence of plastids in the generative cell during development (Hagemann, 1976, 1979). Whether plastids are exculded during generative cell formation, or degenerate later during development (Clauhs and Grun, 1977), their presence or absence may be genetically and taxonomically significant.

Fertilization

Condition of sperm cells at gametic deposition

The male gametes are deposited extremely precisely by the pollen tube through the synergid of the female gametophyte in conventionally-organized angiosperms. Male gametes are present only briefly within the female gametophyte before they are incorporated into either the egg or proendospermatic central cell and therefore, in only three plants are electron microscopic descriptions available: *Hordeum vulgare* (barley, Cass and Jensen, 1970; Cass, 1981; Mogensen, 1982), *Lycopersicon esculentum* (tomato, Kadej and Kadej, 1985), and *Spinacia oleracea* (spinach, Wilms, 1981). Each of these plants displays the removal of the inner pollen plasma membrane prior to fusion in such a manner that the sperm cells are delimited only by their external, wall-less cell membrane. This removal of the delimiting pollen plasma membrane, an event termed analogous to capacitation by Russell (1983), is common to all angiosperms studied to date and is likely to be universal.

The precise location of free, unfused sperm cells is controversial. According to Russell (1983), previous studies have universally shown that the synergid that receives the pollen tube is not intact as it has degenerated to the extent that it lacks a delimiting plasma membrane.

As viewed in this respect, the synergid may merely provide access to interfaces between the egg and proendospermatic cell (cf. Jensen and Fisher, 1968a; Fisher and Jensen, 1969; van Went, 1970). Such interactions have been directly viewed when the free gametes are observed prior to fertilization (Cass and Jensen, 1970; Cass, 1981; Wilms, 1981; Mogensen, 1982; Kadej and Kadej, 1985). The typical cellular interactions present at gametic deposition in a synergid-bearing angiosperm are illustrated in Fig. 7. A review of the condition of the synergid in conventionally-organized megagametophytes after fertilization corroborates this conclusion (Russell, 1983, 1984).

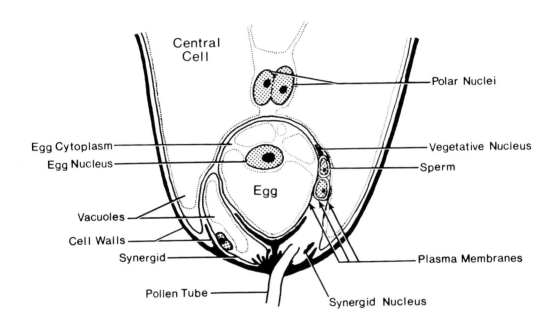

Fig. 7 - Reconstruction of fertilization events in conventionally-organized female gametophyte of an angiosperm. Sperm cells in the female gametophyte are delivered into a degenerate synergid, which at sperm deposition lacks a plasma membrane. As the sperm cells lack an external pollen plasma membrane, the sperm cell membranes are tightly appressed simultaneously to both the egg and central cell. (After Dumas, *et al.*, 1984).

Ultrastructural modifications to organelles

Gametic recognition is a logical prerequisite of fertilization in *Plumbago* and may occur in other angiosperms. Such an event would (1) recognize that a specific cell is a male gamete, and (2) perhaps recognize differences in male gametes, should these occur. If gametic recognition is a feature of other systems, it would occur at gametic deposition. Were the two sperm cells distinguishable, in theory, the following logical prerequisites for successful recognition should be met:

(1) Sperm cell plasma membranes must be directly exposed without an intervening cell wall or pollen plasma membrane, and

(2) Sperm cell membranes contact both the egg and proendospermatic central cell simultaneously.

These two prerequisites might allow even minute differences to be distinguished in a system in which fertilization is directed.

Cytoplasmic heterospermy

Dimorphism in the two sperm cells present within the pollen grain or tube has been infrequently reported in the light microscopic literature. Such dimorphism may include simply physical differences in the sperm cells, differences in cytoplasmic organelles or quantities of organelles, or difference in the nuclear genetics of the sperm cells. Cytoplasmic heterospermy is narrowly defined as a condition in which sperm cells arising from a common generative cell consistently differ in the quantity or quality of heritable organelles. Probably, as more sperm cells are examined, more will be found to possess cytoplasmic heterospermy.

In a broader sense, a number of cases of sperm cells differing in size alone have been reported in the early light microscopic literature, dating to the discovery of double fertilization *Scilla non-scripta*, (Guignard, 1899). Other reports include sperm size differences in *Lilium auriculata* (Blackman and Welsford, 1913), *Fritillaria pudica* (Sax, 1916, 1918), *Iris versicolor* (Sawyer, 1917), *Trillium grandiflorum* (Nothnagel, 1918), *Vallisineria americana* (Wylie, 1923), *Orobanche cumana* (Persidsky, 1926), *Vinca major* (Finn, 1928), *Orobanche ramosa* (Finn and Rudenko, 1930), *Acacia baileyana* (Newman, 1934), *Camassia leichtilinii* (Smith, 1942), and *Cardiospermum halicacabum* (Kadry, 1946).

In the narrow sense, the first description of cytoplasmic heterospermy in which differences in heritable cytoplasmic organelles occur is that of

Plumbago (Russell, 1981, 1984b), in which the number of organelles present in the paired sperm cells was painstakingly described from serial sections using transmission electron microscopy. The organizing structural feature in this dimorphism appears to the presence of a cellular projection that one sperm cell possesses, which physically associates that sperm cell with the vegetative nucleus (Russell and Cass, 1981).

More recently, preferential fertilization has been described in *Plumbago* (Russell, 1985b). This represents only the second convincing case of preferential fertilization (Russell and Cass, 1981; Russell, 1984b) 1985b). Cytoplasmic heterospermy as seen to date, orginates with the presence of a physical association between the vegetative nucleus and one sperm cell and has been described elsewhere in *Brassica oleracea* Dumas, et al., 1985), *B. campestris* (McConchie, *et al.*, 1985), *Hippeastrum vitatum* (Mogensen, 1986), *Petunia* (Wagner and Mogensen, personal communication), and *Spinacia oleracea* (Wilms and van Aelst, 1983). Such physical associations have been shown but not discussed between the vegetative nucleus and generative cell in *Gossypium hirsutum* (Jensen and Fisher, 1970), *Nicotiana tabacum* and *Prunus avium* (Cresti, Ciampolini, and Kapil, 1984), although these associations are likely the origin for later associations between the sperm cell and vegetative nucleus. The absence of such physical association has been confirmed only among the grasses, as described in *Alopecurus pratensis* (Heslop-Harrison and Heslop-Harrison, 1984), *Hordeum vulgare* (Mogensen and Rusche, 1985) and *Triticale* (Schroder, 1983), where such external dimorphism appears not to occur.

In the early light microscopic literature, the smaller of the two sperm nuclei was reported to fuse with the egg in the following flowering plants: *Scilla*, (Guignard, 1899), *Lilium* (Blackman and Welsford, 1913), *Fritillaria* (Sax, 1916, 1918), *Iris* (Sawyer, 1917), *Trillium* (Nothnage, 1918), *Acacia* (Newman, 1934), and *Camassia leichtlinii* (Smith, 1942). The larger of the two sperm nuclei was reported to fuse with the egg in *Orobanche cumana* (Persidsky, 1926) and *Orobanche ramosa* (Finn and Rudenko, 1930). Other morphological variants have also been reported; for example, in *Cardiospermum halicacabum,* the male gametic nucleus that fertilizes the egg is round and swollen, tapering into a long narrow nucleus at its opposite end (Kadry, 1946), whereas the other is nondescript.

Ultrastructural modifications to organelles

In *P. zeylanica*, the sperm more likely to fertilize the egg cell, the Sua, is the smaller of the two sperm cells (29.6% smaller than the Svn in cellular volume) with a nucleus 39.2% smaller in nuclear volume than the Svn. This corresponds to a difference in nuclear diameter of 15.2% assuming that the nucleus is spherical and therefore may still be detectable at light microscopic resolution level, (Weibel, 1979). As Maheshwari (1950) notes, the conclusions derived from these studies are subject to sectioning artifacts and orientation of the nucleus. With modern techniques, the naturally-occurring differences in nuclear size, if examined soon after gametic fusion, might be used to more easily trace sperm cell fate than tracing the fate of sperm cytoplasmic organelles as in *Plumbago* (Russell, 1985b). Since sperm nuclear size can be more accurately measure using modern techniques than those used by early researchers, such a directly observable criterion could be useful in determining the fate of sperm cells displaying cytoplasmic heteromorphism.

Nuclear heterospermy

Nuclear heterospermy, a condition in which the sperm cells differ in their nuclear genetic complement, can arise only through the non-disjunction of genetic elements during division of the generative cell. Such non-complementarity can arise either through mistakes in segregation, incorrect replication or non-replication. Presumably, any foreign or naturally-occurring nuclear DNA that does not segregate in a Mendelian fashion may result in nuclear heterospermy. To date, the most clearly described case of nuclear heterospermy is that in B-chromosomes of certain strains of maize. Non-disjunction at generative cell division occurs as much as 80% of the time in this plant and therefore may be an important factor in sexual reproduction in this plant. In order to determine the pattern of gametic fusion during fertilization, genetic markers occurring exclusively on the B-chromosomes were used to trace the fate of the sperm cells (Roman, 1948). In 76.8% of the non-disjunct B^4 chromosomes in the TB-4a strain and 67.1% of the non-disjunct B^9 chromosomes in the TB-9b strain, the hyperploid sperm cell, with a duplicate B-chromosome, was transmitted into the egg cell to form hyperploid embryos. Roman (1948) concluded that sperm cells with an extra B-chromosome had participated in "directed fertilization," in his terminology.

102

Two hypotheses for this behaviour were proposed: (1) genetic influences from the B-chromosome affected the pattern of fusion, or (2) the sperm cells are polarized in organization or behaviour. Roman discarded the first hypothesis since it seems highly unlikely that B-chromosomes arising from two different chromosomes could each carry a genetic factor favouring its own transmission during fertilization (Roman, 1948). He argued instead that preferential fusion may result from a preferred pattern of sperm cell arrival, with the first sperm cell discharged into the female gametophyte having an advantage in selecting the pattern of fusion (Roman, 1948). Gametic deposition, however, is likely to be as rapid in vivo as is pollen tube burst in vitro, and therefore too rapid to confer a temporal advantage to either sperm cell (Russell, 1985b). An alternative hypothesis is that the sperm cells differ from their inception, but that these differences are revealed only because a preferential behaviour of the mitotic spindle at generative cell mitosis results in the transmission of non-pairing chromosomes into the sperm cell more likely to fuse with the egg Russell, 1985b).

It appears that the non-disjunction of the B-chromosome at generative cell division therefore may allow differences in the fertilization capacity of the sperm cells to be seen without itself directly predisposing the sperm cells to preferential fertilization (Russell, 1985b). Preliminary ultrastructural work using corn has failed to demonstrate dimorphism in sperm cells (McConchie, *et al.*, personal communication), although the capacity for preferential fusion may be present in this plant (Roman, 1948). Thus, the capacity for preferential fertilization apparently may be present in these sperm cells from their inception whether or not such differences are readily evident.

Mechanism of gametic fusion

Although Maheshwari (1950) notes that very few detailed accounts of gametic fusion prior to double fertilization had appeared as of his publication date, the situation has not changed greatly since that time. Among the most detailed examinations of that era were the works of Anderson (1939) and Wylie (1923, 1941). These studies demonstrated a retention of the male cytoplasm until gametic fusion and strongly

suggested a plasmatic nature to gametic fusion. The later work of
Wilms (1981) and Mogensen (1982) have confirmed through ultrastructural
observations that sperm cells remain intact prior to fusion, but differ in
their mechanisms of fusion. Wilms (1981) proposed after a single
observation of a two nucleated sperm cell between the egg and central cell
that the two sperm cells may fuse first with one another and then with the
egg and central cell, with an elimination of organelles. In the absence
of confirming data from either light or electron microscopy, this
observation and theory remains enigmatic. Mogensen (1982) reported
that a large enucleated cytoplasmic body roughly corresponding in size
to sperm cytoplasm may occur near the synergid soon after gametic fusion
and proposed, similarly to Jensen and Fisher (1968a), that the sperm
cytoplasm is shed during fusion with the egg. The exact mechanism by
which a sperm nucleus could be transmitted into female reproductive cells
without cytoplasm accompanying the nucleus still remains elusive in the
absence of more detailed data. If fusion is initiated as a plasmatic
event between male and female cell membranes (Jensen, 1974; Mogensen,
1982), the reclosure of the fusion suture without transmission of
cytoplasmic organelles is not easily explained. Current studies of fusion
in other biological systems fail to provide a mechanism for organelle
exclusion by this mechanism.

In *Plumbago*, there is complete transmission of the cellular material
present in the sperm cell at the time of fusion. Any enucleated
cytoplasmic bodies arising from the sperm cells occur by severance of sperm
cell cytoplasm at the time of expulsion from the pollen tube (Russell,
1983). By this mechanism, an estimated 15% of the sperm mitochondria are
lost from the fusion process. Mogensen and Rusche (1985) describe an
organellar diminution scheme for maturing barley pollen grains in which long
protuberances of the sperm cells form, sever, and are reassimilated into
the vegetative cytoplasm in a loss of nearly 50% of the mitochondria of
barley sperm before anthesis. Such an organelle diminution scheme could
continue in operation within the pollen tube, give rise to such bodies and,
were these to accumulate or fuse in one region, could form a large body
similar to the one described between the egg and central cell (Mogensen,
1982), without contradicting the *Plumbago* model (Russell, 1983). In
the 1/3 of described plants possessing biparental inheritance of plastids

(Grun, 1976; Gillham, 1978; Kirk and Tilney-Bassett, 1978), clearly the *Plumbago* model is the most appropriate.

Light and electron micrographs of the transient phase in which the sperm cells are present in the synergid prior to gametic fusion are rare, but available (Wylie, 1941; Cass and Jensen, 1970; Cass, 1981; Wilms, 1981; Mogensen, 1982; Kadej and Kadej, 1985) and meet the predictions from *Plumbago* results that the sperms actually exit the chalazal end of the synergid, thus gaining access to a space between the egg and central cell that is homologous to that in synergid-less angiosperms (Fig. 7; see also Russell, 1983, 1984a). With the two sperm cells in direct, simultaneous contact with both the egg and central cell (Fig. 7), the structural prerequisites for gametic recognition are present in the majority of flowering plants. At this point, cellular modes of recognition as seen in other plant systems (van de Ende, 1976) may become important and serve as models in understanding fertilization events in angiosperms. The occurrence of simultaneous contacts illustrated herein (Fig. 7) may be important for gamete-level recognition and may occur in numerous angiosperms. These schemes are supported by data presented in cotton (Fisher and Jensen, 1969), barley (Cass and Jensen, 1970; Mogensen, 1982, tomato (Kadej and Kadej, 1985), and spinach (Wilms, 1981). To date, there have been no reports to contradict it.

Impact of transmitted male cytoplasm

Undeniably, the most compelling evidence of male cytoplasmic inheritance are the numerous examples of biparental plastid inheritance. In these cases, clearly the plastids are directly transmitted into the egg through a plasmatic form of gametic fusion (Meyer and Stubbe, 1974; Russell, 1983). Both reports emphasize that the ultrastructural characteristics are sperm plastids and not pollen tube plastids, which are morphologically different (Meyer and Stubbe, 1974; Russell, 1983). According to absolute plastid counts in *Oenothera erythrosepala*, a plant known to inherit plastids biparentally, the young zygote contained an average of 27.3 egg plastids and 10.7 sperm plastids (Meyer and Stubbe, 1974). Egg plastids also larger in volume, an average of $5.16 \mu m^3$, compared to an average of $0.67 \mu m^3$ for sperm plastids.

Despite these numerical iniquities, sperm plastids are not eliminated from expression in the resulting offspring and their derivatives can easily be followed externally by the use of color mutants (Grun, 1976; Gillham, 1978). In *Plumbago*, preliminary data suggests that the number of plastids present in the egg is approximately 110 prior to fusion, occupying an average profile area of $0.853\mu m^2$ (Tables II, V). The sperm cell that fertilizes the egg 94% of the time contains an average of 24.3 plastids, occupying an average profile area of 0.518 μm^2 (Tables I, V) Although no studies of the plastid genetics of any members of the *Plumbaginaceae* have emerged, the ratios of sperm to egg plastid numbers and volumes appear comparable. Based on his data, biparental inheritance of plastids would be predicted in *Plumbago*..

The corresponding numerical data concerning mitochondrial inheritance suggests that the contribution made by male mitochondria into the zygote is minimal at best and that it may be highly disadvantaged in its expression (Pearson, 1981). In *Plumbago*, the sperm cell that is typically successful in fertilizing the egg contains an average of 39.8 mitochondria (Table I) with each mitochondrion occupying $0.26\mu m^3$. The egg cell, however, by preliminary estimate, contains approximately 14,000 mitochondria (Table II) with each mitochondrion occupying $0.48\mu m$. The smaller size of the mitochondria may, as suggested by Pearson (1981), contribute to disadvantaging the expression of the male mitochondrial genome in the mature plant. Clearly, the wide discrepancy in numbers of male versus female mitochondria would disadvantage expression of the male mitochondria. Emerging data on other plants, although only reporting the numbers of mitochondria in the sperm cells, shows a similar pattern in other flowering plants. In *Brassica campestris*, the Svn has an average of 23.4 mitochondria and the Sua has an average of 6.4 mitochondria (McConchie, *et al.*, 1986). In *Brassica oleracea*, the Svn has an average of 12.7 mitochondria and the Sua has an average of 8.7 mitochondria (McConchie, *et al.*, 1986). *Hordeum vulgare,* in the mature sperm cell, contains an average of 30 mitochondria in its isomorphic cells (Mogensen and Rusche, 1985). In spinach, the number of mitochondria within the sperm cells is also, on average, less that 25 (Wilms, personal communication). Similarly, low numbers of mitochondria are present in corn (McConchie, *et al.*, personal communication).

Mitochondria appear to be at a numerical disadvantage from the formation of the generative cell, but during development there is a mechanism present

in barley that further diminishes the number of heritable organelles in
sperm cells. According to Mogenson and Rusche (1985), vesicles containing
organelles may bud off of the sperm cells and degenerate without a subsequent replacement of these organelles during development. In fact, the
mitochondria decrease in volume during this period in development,
reaching a lower value at maturity than earlier in development (Mogensen
and Rusche, 1985).

The evolutionary advantages and disadvantages of uniparental versus
biparental inheritance of plastids and mitochondria are at best poorly
understood (Grun, 1976; Gillham, 1978). There is little evidence for
populational competition for superior genotypes within the mt-DNA and
ct-DNA of a higher plant as such competition would occur on a cellular
level where some metabolic needs may be met by surrounding cells (Birky,
1983). For instance, albino plastids do not outcompete green plastids
within the cell; rather, the two remain at equilibrium (Birky, 1983).
The chances for complete fixation of heterogeneous populations decreases
with the number of particles and for more than just several such particles,
the number of mitotic cycles required for fixation of the organelle type
rapidly reaches and then exceeds the number of mitotic cycles in a plant's
lifetime (Grun, 1976). The seeming disadvantages of uniparental inheritance,
including the possibility that a pollen plant with an inferior plastid
genotype may be able to compete evenly with a better plastid genotype,
have been suggested (Grun, 1976), but still seem to be contradicted by the
fact that the number of plants that are successful with biparental inheritance is limited. Somehow, it seems unlikely that an unsuccessful plant
would have sufficient reproductive success to merit such a protective
mechanism.

A more important rationale for evolutionary advantages of maternal
cytoplasmic inheritance may be that numerous organellar proteins are
assembled from both nuclear-encoded polypeptide segments or proteins and
those encoded by organellar DNA. In those cases where two or more
polypeptides of different origins within the cell combine in order to
produce a functional unit, any significant change in protein structure of
either subunit may impair the function of the resulting enzyme. As the
maternal genotype of both the nucleus and cytoplasm are retained in the
offspring, there may be a greater probability of creating functional enzymes.

Ultrastructural modifications to organelles

   Mitochondria have such a strong disadvantage in transmission, that it is possible that some mitochondrial traits expressed during pollen tube growth resulting in rapid growth of the pollen tube may not be beneficial to the mature plant. These characteristics may include greater tolerance to anaerobic conditions and specific exploitative attributes in pollen competition that do not contribute to or are deleterious to cooperative cell function. Such functions may help a specific pollen tube to arrive at the egg prior to its competitors, but once transmitted, may be sufficiently impaired in more normal embryonic or post-embryonic function that it stunts the offspring. The poorly understood nature of mitochondrial transmission in angiosperms stems from a lack of good genetic markers for the mitochondrion (Grun, 1976; Gillham, 1978). All conventional mitochondri markers that could be easily traced by geneticists seem to influence mitochondrial function. Until paternal mitochondrial transmission is fully understood, the reasons for its apparent absence will remain enigmatic.

CONCLUSIONS

Using the above as a comparative basis, it is possible to view the present results in their appropriate perspective. A number of the results reported here have evident importance: (1) materials introduced into the sperm cytoplasm are transmitted into the egg, (2) evidence indicates that avenues for the transmission of pollen-borne virions coincide with the intercellular spaces described in the previous section and (3) gametic recognition phenomena could be used to specifically target certain materials into either the egg or central cell for future expression in the embryo or endosperm (Russell, 1984a). Model systems such as *Plumbago* may provide a perspective from which to approach the salient problems of reproduction, which may then be answered more analytically in other, more important flowering plant species.

ACKNOWLEDGEMENTS

I wish to thank Susan M. Heinrichs, Norman R. Geltz, and Timothy W. Mislan for technical assistance. Support for the research reported here was provided in part by U.S. National Science Foundation grants PCM-8-08466 and PCM-8409151.

REFERENCES

Allington, P.M. (1985). Micromanipulation of the unfixed cereal embryo sac
in *Experimental manipulation of ovule tissues*. (eds. Chapman, G.P.,
Mantell, S.H., and Daniels, R.W.) pp. 39 - 51, Longman.

Anderson, L.E. (1939). Cytoplasmic inclusions in the male gametes of
Lilium. *Amer. J. Bot. 26*, 761 - 6.

Angold, R.E. (1967). The ontogeny and fine structure of the pollen grain
of *Endymion non-scriptus*. *Rev. Paleobot. Palynol. 3*, 205 - 12.

Baur, E. (1909). Das Wesen und die Erblichkeitsverhaltnisse der
"varietates albomarginatae hort" von *Pelargonium zonale*. *Z.Vererbunqs.*
1, 330 - 51.

Birky, C.W. (1978). Transmission genetics of mitochondria and chloroplasts.
Ann. Rev. Genet. 12, 471 - 512.

Birky, C.W. (1983). Relaxed cellular controls and organelle heredity.
Science 222, 468 - 75.

Blackman, V.H., and Welsford, E.J. (1913). Fertilisation in *Lilium*. *Ann.*
Bot. 27, 111 - 114.

Bopp-Hassenkamp, G. (1960). Elektronenmikroskopiche Untersuchungen an
Pollenschlauchen zweier *Liliaceen*. *Z. Naturforsch. 15b*, 91 - 4.

Burgess, J. (1970). Cell shape and mitotic spindle formation in the
generative cell of *Endymion non-scriptus*. *Planta 95*, 72 - 85.

Cass, D.D. (1973). An ultrastructural and Normarski-interference study of
the sperms of barley. *Can. J. Bot. 51*, 601 - 5.

Cass, D.D. (1981). Structural relationships among central cell and egg
apparatus cells of barley as related to transmission of male gametes.
Acta Bot. Soc. Pol. 50, 177 - 9.

Cass, D.D., and Karas, I. (1974). Ultrastructural organization of the egg
of *Plumbago zeylanica*. *Protoplasma 81*, 49 - 62.

Cass, D.D., and Karas, I. (1975). Development of the sperms of barley.
Can. J. Bot. 53, 1051 - 62.

Cass, D.D., and Jensen, W.A. (1970). Fertilization in barley. *Amer.*
J. Bot. 57, 62 - 70.

Clarke, A.E., and Knox, R.B. (1978). Cell recognition in flowering plants.
Q. Rev. Biol. 53, 3 - 28.

Caluhs, R.P., and Grun, P. (1977). Changes in plastid and mitochondrion content during maturation of generative cells of *Solanum (Solanaceae)*. *Amer. J. Bot. 64*, 377 - 83.

Cocucci, A., and Jensen, W.A. (1969). Orchid embryology: pollen tetrads of *Epidendrum scutella* in the anther and on the stigma. *Planta 84*, 215 - 29.

Correns, C. (1909). Vererbungsversuche mit blab (gelb) grunen und buntblattrigen Sippen bei *Mirabilis jalapa, Urtica pilifera und Lunaria annua. Z. Vererbungs. 1*, 291 - 329.

Cresti, M., Ciampolini, F., and Kapil, R.M. (1984). Generative cells of some angiosperms with particular emphasis on their microtubules. *J. Submicrosc. Cytol. 16*. 317 - 26.

Dahlgren, K.V.O. (1916). Zytologische und embryologische Studien uber die Reihen Primulales und Plumbaginales. *K. Svensk. Ventenskap. Handl. 56*, 1 - 80.

Dahlgren, K.V.O. (1937). Die Entwicklung des *Embryosackes bei Plumbago zeylanica. Bot. Notiser 1937*, 487 - 498.

daSilva, P.P., and Kachar, B. (1980). Quick-freezing versus chemical fixation: capture and identification of membrane fusion intermediates. *Cell Biol. Intl. Rep. 4*, 625 - 40.

Day, A., and Ellis, T.H.N. (1984). Chloroplast DNA deletions associated with wheat plants regenerated from pollen: possible basis for maternal inheritance of chloroplasts. *Cell 39*, 359 - 68.

Dexheimer, J. (1965). Sur les structures cytoplasmique dans les grains de pollen dans *Lobella erinus. C.R. Acad. Sci. Paris. Ser. D. 260*, 6963 - 5.

Diers, L. (1963). Elektronenmikroskopische Beobachtungen an der generativen Zellen von *Oenothera hookeri. Z. Naturforsch. 18*, 562 - 6.

Dumas, C., Knox, R.P., and Gaude, T. (1985). The spatial association of the sperm cells and vegetative nucleus in the pollen grain of *Brassica*. *Protoplasma 124*, 168 - 74.

Dumas, C., Knox, R.B., McConchie, C.A., and Russell, S.D. (1984). Emerging physiological concepts in fertilization. *What's New Plant Physiol. 15*, 17 - 20.

Dupuis, F. (1972). Formation of wall between vegetative and generative cells of pollen from *Impatiens balsamina* L. *Bull. Soc. Bot. Fr. 119*, 41 - 50.

Finn, W.W. (1928). Spermazelle bei *Vinca major und V. herbacea. Ber Dtsch. Bot. Gesell. 46,* 235 - 46.

Finn, W.W., and Rudenko, T. (1930). Spermatogenesis und Befruchtung bei einigen *Orobanchaceae. Bull. Jard. Bot. de Kieff. 11,* 69 - 82.

Fisher, D.B., and Jensen, W.A. (1969). Cotton embryogenesis: the identification, as nuclei, of the X-bodies in the degenerated synergid. *Planta 84,* 122 - 33.

Gillham, N.W. (1978). *Organelle heredity.* Raven Press.

Gorska-Brylass, A. (1970). The callose stage of the generative cells in pollen grains. *Grana 10,* 21 - 30.

Grun, P. (1976). *Cytoplasmic genetics and evolution.* Columbia University Press.

Guignard, L. (1899). Sur les antherozoides et la double copulation sexualle chez les vegetaux angiospermes. *C. R. Acad. Sci. Paris 128,* 864 - 71.

Hagemann, R. (1976). Plastid distribution and plastid competition in higher plants and the induction of plastom mutations by nitroso-urea compounds in *Genetics and biogenesis of chloroplasts and mitochondria.* (eds. Bucher, T., Neupert, W., Sebald, W., and Werner, S.) pp. 331 - 8, North-Holland.

Hagemann, R. (1979). Genetics and molecular biology of plastids of higher plants. *Stadler Symp. 11,* 91 - 115.

Heslop-Harrison, J. (1968). Synchronous pollen mitosis and the formation of the generative cell in masculate orchids. *J. Cell Sci. 3,* 457 - 66.

Heslop-Harrison, J., and Heslop-Harrison, Y. (1984). The disposition of gamete and vegetative cell nuclei in the extending pollen tube of a grass species, *Alopecurus pratensis. Acta Bot. Neerl. 33,* 131 - 4.

Hoefert, L.L. (1968). Ultrastructure of *Beta* pollen. *1.* Cytoplasmic constituents. *Amer. J. Bot. 56,* 363 - 8.

Hoefert, L.L. (1969). Fine structure of sperm cells in pollen grains of *Beta. Protoplasma 68,* 237 - 40.

Hoefert, L.L. (1971). Pollen grain and sperm cell ultrastructure in *Beta.* in *Pollen: development and physiology.* (eds. Heslop-Harrison, J.) pp. 68 - 9, Appleton-Century-Crofts.

Horvat, F. (1969). La paroi de la cellule generative du grain de pollen. *Pollen Spores 11,* 181 - 201.

Jensen, W.A. (1974). Reproduction in flowering plants in *Dynamic aspects of plant ultrastructure* (ed. Robards, A.W.) pp. 481 - 503, McGraw-Hill.

Jensen, W.A., and Fisher, D.B. (1968a). Cotton embryogenesis: the entrance and discharge of the pollen tube in the embryo sac. *Planta 78,* 158 - 83.

Jensen, W.A., and Fisher, D.B. (1968b). Cotton embryogenesis: the sperm. *Protoplasma 65,* 277 - 86.

Jensen, W.A., and Fisher, D.B. (1970). Cotton embryogenesis: the pollen tube in the stigma and style. *Protoplasma 69,* 215 - 35.

Kadej, A., and Kadej, F. (1985). Sperm cells in the fertilization in *Lycopersicon esculentum* in *Sexual reproduction in seed plants, ferns and mosses.* (eds. Willemse, M.T.M., van Went, J.L.) pp. 149, Pudoc.

Kadry, A.E.R., (1946). *Embryology of Cardiospermum halicacabum L. Svensk Bot. Tidskr. 40,* 111 - 26.

Karas, I., and Cass, D.D. (1976). Ultrastructural aspects of sperm cell formation in rye: evidence for cell plate involvement in generative cell division. *Phytomorphology 26,* 36 - 45.

Kirk, J.T.O., and Tilney-Bassett, R.A.E. (1978). *The plastids: their chemistry, structure, growth and inheritance.* Elsevier Biomedical Press, Amsterdam.

Larsen, D.A. (1963). Cytoplasmic dimorphism within pollen grains. *Nature 200,* 911 - 2.

Larsen, D.A. (1965). Fine structural changes in the cytoplasma of germinating pollen. *Amer. J. Bot. 52,* 139 - 55.

Lutz, R.W., and Sjolund, R.D., (1973). Development of the generative cell wall in *Monotropa uniflora L. pollen. Plant Physiol. 52,* 498 - 500.

Maheshwari, P. (1950). *An introduction to the embryology of angiosperm.* McGraw-Hill.

Maruyama, K., Gay, H., and Kaufmann, B.P. (1965). The nature of the wall between generative cell and vegetative nuclei in the pollen grain of *Tra-escantia paludosa. Amer. J. Bot. 52,* 605 - 10.

McConchie, C.A., Jobson S., and Knox, R.B. (1985). Computer-assisted reconstruction of the male germ unit in pollen of *Brassica campestris. Protoplasma 127,* 57 - 63.

McConchie, C.A., Dumas, C., Jobson, S., Russell, S.D., and Knox, R.B. (1986). Quantitative cytology of the sperm cells of *Brassica campestris and Brassica oleracea*. *Planta* (In Preparation).

Meyer, B., and Stubbe, W. (1974). Das Zahlenverhaltnis von mutterlichen und vaterlichen Plastiden in den Zygoten von *Oenothera erythrosepala Borbas* (syn. *Oe. lamarckiana*). *Ber. Dtsch. Bot. Ges. 87*, 29 - 38.

Mogensen, H.L. (1982). Double fertilization in barley and the cytological explanation for haploid embryo formation, embryoless caryopses, and ovule abortion. *Carlsberg Res. Comm. 47*, 313 - 54.

Mogensen, H.L. (1986). On the male germ unit in an angiosperm with bicellular pollen, *Hippeastrum vitatum* in *Pollen biotechnology and ecology*. (eds. Mulcahy, D.L., and Ottaviano, E.) pp. Springer Verlag.

Mogensen, H.L., and Rusche, M.L. (1985). Quantitative analysis of barley sperm: occurrence and mechanism of cytoplasm and organelle reduction and the question of sperm dimorphism. *Protoplasma 128*, 1 - 13.

Newman, I.V. (1934). Studies in the Australian acacias. IV. The life history of *Acacia baileyana* F.V.M. Part 2. Gametophytes, fertilization, seed production and germination, and general conclusions. *Proc. Linn. Soc. N.S. Wales 59*, 277 - 313.

Nothnagel, M. (1918). Fecundation and formation of the primary endosperm nucleus in certain *Liliaceae*. *Bot. Gaz. 66*, 141 - 61.

Owens, S.J., and Westmuckett, A.D. (1983). The structure and development of the generative cell wall in *Gibasis karwinskyana, G. Venustula and Tradescantia blossfeldiana (Commelinaceae)* in *Pollen: biology and implication for plant breeding*. (eds. Mulcahy, D.L., and Ottaviano, E.) pp. 149 - 57, Elsevier Biomedical Press.

Pearson, G.H. (1981). Nature and mechanism of cytoplasmic male sterility in plants: a review. *Hort Science 16*, 482 - 7.

Persidsky, D. (1926). Zur Embryologie der *Orobanche cumana* Wall. und der *O. ramosa* L. *Bull. Jard. Bot. de Kieff. 4*, 6 - 10.

Roland, F. (1971). Characterization and extraction of polysaccharides of the intine and of the generative cell wall in pollen grains of some *Ranunculaceae*. *Grana 11*, 101 - 6.

Roman, H. (1948). Directed fertilization in maize. *Proc. Natl. Acad. Sci. (USA) 34*, 36 - 42.

Russell, S.D. (1980). Participation of male cytoplasm during gamete fusion in an angiosperm, *Plumbago zeylanica*. *Science 210,* 200 - 1.

Russell, S.D. (1981). Fertilization in *Plumbago zeylanica:* the structural basis of male cytoplasmic inheritance. Ph.D. dissertation, University of Alberta, Edmonton, Canada.

Russell, S.D. (1982). Fertilization in *Plumbago zeylanica:* entry and discharge of the pollen tube into the embryo sac. *Can. J. Bot. 60,* 2219 - 30.

Russell, S.D. (1983). Fertilization in *Plumbago zeylanica:* gametic fusion and fate of the male cytoplasm. *Amer. J. Bot. 70,* 416 - 34.

Russell, S.D. (1984a). Timetable of fertilization in angiosperms in *Pollination '84* (eds. Williams, E., and Knox, R.B.) pp. 69 - 77, School of Botany, University of Melbourne.

Russell, S.D. (1984b). Ultrastructure of the sperm of *Plumbago zeylanica:* 2. Quantitative cytology and thre-dimensional reconstruction. *Planta 162,* 385 - 91.

Russell, S.D. (1985a). Microgametogenesis in *Plumbago zeylanica. Amer. J. Bot. 72,* 830 - 1.

Russell, S.D. (1985b). Preferential fertilization in *Plumbago:* ultrastructural evidence for gamete-level recognition in an angiosperm. *Proc. Matl. Acad. Soc. (USA) 82,* 6129 - 32.

Russell, S.D. (1986). The isolation of sperm of *Plumbago zeylanica. Plant Physiol. 81,* 317 - 19.

Russell, S.D., and Cass, D.D. (1981). Ultrastructure of the sperm of *Plumbago zeylanica:* 1. Cytology and association with the vegetative nucleus. *Protoplasma 107,* 85 - 107.

Sanger, J.M., and Jackson, W.T., (1971a). Fine structure study of pollen development in *Haemanthus katherinae* Baker. I. Formation of vegetative and generative cells. *J. Cell Sci. 8,* 289 - 301.

Sanger, J.M., and Jackson, W.T., (1971b). Fine structure study of pollen development in *Haemanthus katherinae* Baker. II. Microtubules and elongation of the generative cells. *J. Cell Sci. 8,* 303 - 15.

Sanger, J.M., and Jackson, W.T. (1971c). Fine structure study of pollen development in *Haemanthus katherinae* Baker. III. Changes in organelles during development of the vegetative cells. *J. Cell Sci. 8,* 317 - 29.

Sassen, M.M.A. (1964). Fine structure of *Petunia* pollen grain and pollen tube. *Acta Bot. Neerl. 13,* 175 - 81..

Sassen, M.M.A., and Kroh, M. (1974). The "cell wall" around the generative cell. *Acta Bot. Neerl. 23,* 354 - 5.

Sawyer, M.L. (1917). Pollen tubes and spermatogenesis in *Iris*. *Bot. Gaz. 64,* 159 - 64.

Sax, K. (1916). Fertilization in *Fritillaria pudica*. *Bull. Torrey Bot. Club 43,* 505 - 22.

Sax, K. (1918). The behaviour of the chromosomes in fertilization. *Genetics 3,* 309 - 27.

Schroder, M.B. (1983). The ultrastructure of sperm cells in *Triticale* in *Fertilization and Embryogenesis in Ovulated Plants* (ed. Erdelska, O.) pp. 101 - 4, Veda.

Smith, F.H. (1942). Development of the gametophytes and fertilization in *Camassia. Amer. J. Bot. 29,* 657 - 63.

Thiery, J.P. (1967). Mise en evidence des polysaccharides sur coupes fines en microscopie electronique. *J. Microsc. 6,* 987 - 1018.

van den Ende, J. (1976). *Sexual interactions in lower plants.* Academic Press.

Vaughn, K.C., Debronte, L.R., Wilson, K.G., and Schaeffer, G.W. (1980). Organelle alteration as a mechanism for maternal inheritance. *Science 208,* 196 - 198.

Weibel, E.R. (1979). *Stereological methods. Vol. 1. Practical methods for biological morphometry.* Academic Press.

Went, J.L. van (1970). The ultrastructure of the fertilized embryo sac of *Petunia. Acta Bot. Neerl. 19,* 468 - 80.

Wilms, H.J. (1981). Pollen tube penetration and fertilization in spinach. *Acta Bot. Neerl. 19,* 101 - 22.

Wilms, H.J., and van Aelst, A.C. (1983). Ultrastructure of sperm cells in mature pollen in *Fertilization and Embryogenesis in Ovulated Plants* (eds. Erdelska, O.) pp. 105 - 12, Veda.

Wylie, R.B. (1923). Sperms of *Vallisneria spirilis. Bot. Gaz. 75,* 191 - 202.

Wylie, R.B. (1941). Some aspects of fertilization in *Vallisneria. Amer. J. Bot. 28,* 169 - 74.

Ultrastructural modifications to organelles

Zhou, C., and Yang, H.Y. (1985). Observations on enzymatically isolated, living and fixed embryo sacs in several angiosperm species. *Planta 165*, 225 - 31.

Zhu, C., Hu, S.V., Xu, L.V., Li, H.R., and Sen, J.H. (1980). Ultrastructure of the sperm cells in mature pollen grains of wheat. *Sci. Sinica 23*, 371 - 9.

Molecular analysis of the chloroplast genomes of plants

5   The *Vicia faba* chloroplast genome

K. Ko

ABSTRACT The organisation of the chloroplast genome is highly conserved among plants. The single most prominent event in the evolution of the chloroplast genome is the deletion of one of the inverted repeated segments in a group of legumes which includes *Vicia faba*. This event may be accompanied by subsequent complex rearrangements; the rearrangements in *V. faba* are limited to two inversions that split the inverted repeated region. The well-defined alterations that occurred to form the chloroplast genome of *V. faba* provide a unique opportunity to study the effects of structural rearrangements on the function of the chloroplast genome. This article reviews the structure, molecular genetics and molecular evolution of the chloroplast genome of *V. faba* and related organisms.

INTRODUCTION

Photosynthesis in green plants is a complex process that requires the coordinated collaboration of two different organelles - the nucleus and the chloroplast. Each of these compartments possess its own genome and the ability to replicate, transcribe and translate its genetic information. The nucleus contributes and controls the largest proportion of the proteins found in the chloroplast; the chloroplast genome, although small in comparison, plays an essential role. The extent of this role has yet to be fully elucidated.

Two dimensional gel analysis revealed the presence of at least 200 separable polypeptides in chloroplasts of *Pisum sativum*. Among these polypeptides, approximately 12 thylakoid membrane polypeptides and at least 80 other polypeptides from the stroma fraction were radiolabelled in *in vitro* chloroplast protein synthesis experiments (Ellis, 1981). Although the chloroplast genome has the capacity to encode 100 or more polypeptide genes, only about one quarter of these genes has been identified

119

and located on the chloroplast DNA (Groot, 1984). Factors that control and regulate the expression of these identified genes are just beginning to be unveiled. Even less is known about the factors that control and regulate gene expression in different plastid types or chloroplasts at different stages of plant development.

The general organisation of the *Vicia faba* chloroplast genome differs greatly from genomes characterised in other higher plants like *Spinacia oleracea*. The differences provide opportunities for studying the effect of structural rearrangements on the function of the chloroplast genome. In this review, various aspects of chloroplast molecular biology pertaining to the *Vicia faba* chloroplast genome will be discussed. It is not intended that it be a comprehensive review of chloroplast molecular genetics; for this the reader is referred to the recent excellent review articles in this field (Palmer, 1985; Bohnert, Crouse and Schmidt, 1982; Bottomley and Bohnert, 1982; Whitfeld and Bottomley, 1983).

PHYSIOCHEMICAL PROPERTIES AND GENERAL STRUCTURE OF THE CHLOROPLAST GENOME

Isolation of Chloroplast DNA

A prerequisite to studies on the structure and function of the chloroplast genome is a method for the purification of chloroplast DNA free from nuclear DNA contamination. Published procedures for the isolation of chloroplast DNA vary significantly for the different plants that have been studied. Unfortunately, methods that work well with some plants may not work for others. Chloroplast DNA from *Vicia faba* can be isolated consistently by the method described by Ko, Straus and Williams (1983). This method involves blending *V. faba* leaves in liquid nitrogen and then dispersing the leaf powder in a buffer containing sorbitol, EDTA, mercaptoethanol, BSA and Tris - HCl pH 8.O. The homogenate is filtered and the chloroplasts collected by centrifugation. Nuclear and cellular contamination is removed by floating the chloroplasts on a 55% buffered - sucrose solution. The layered chloroplasts can then be collected, resuspended and lysed with detergents. Proteinaceous material is removed by extracting the lysate with organic solvents. The chloroplast DNA can then be collected by alcohol precipitation. The DNA isolated by this method can be used directly for analysis without further purification.

The Molecular Structure of the Chloroplast Genome

The physicochemical properties and molecular form of all chloroplast DNAs so far examined are remarkably similar. All chloroplast genomes consist of a covalently closed, double-stranded, circular DNA molecule. The bouyant density of chloroplast DNA's as determined by density gradient ultra-centrifugation in neutral CsCl falls within a range of 1,699 g/cm which converts to a dG/dC content of 37 – 38% (Mandel, Schildkraut and Marmur, 1968) The *Vicia faba* chloroplast genome does not deviate significantly from these values; the bouyant density was calculated to be 1.696 g/cm (Kung and Williams, 1969) which corresponds to 37% in dG/dC content.

The pattern of inheritance of the chloroplast genome is predominantly maternal and mature chloroplasts contain many copies of the chloroplast genome. In spite of the large numbers of genomes per chloroplast, the molecular size of the chloroplast chromosome is uniform both within an individual plant and between different individuals of the same species. The size of most higher plant chloroplast molecules, determined by electron microscopy and by restriction enzyme analysis, is in the range of 130,000 – 155,000 nucleotide base pairs.

The general structure of the chloroplast genome of most vascular plants has been strongly conserved throughout evolution. It contains a sequence, ranging from 22,000 base pairs, repeated once in an inverted configuration, The inverted repeats are separated by two unique sequences. A large single copy sequence (from 78 to 100 kbp) and a small single copy sequence (12 to 30 kbp) separate the two ends of the inverted repeat regions. The few reported changes in chloroplast genome size are generally accounted for by extending or shortening the length of the inverted repeat sequences. In the geranium, *Pelargonium hortorum,* the genome size has increased to 217 kbp by extending the inverted repeat sequences to 76 kbp; on the other hand, in coriander, *Coriandrum sativum,* the inverted repeat sequences are shortened to no more than half the normal size (Palmer, 1985).

Vicia faba is a member of a group of legumes that have lost one of the inverted-repeats. The deletion of one entire segment of the inverted repeat is accompanied by a reduction in the genome size; the total length of the *V. faba* chloroplast as measured by electron microscopy is 39 m which converts to about 121,000 bp (Koller and Delius, 1980). This is approximately 17,000 – 20,000 base pairs smaller than the smallest inverted repeat - containing chloroplast genome.

Restriction Enzyme Analysis of the Chloroplast Genome

Type II restriction endonucleases recognize short DNA sequences and
cleave DNA at these sites. Fragments produced by any given restriction
enzyme digestion can be separated according to their size by gel
electrophoresis. There appears to be a relationship between base
composition and the frequency of specific restriction endonuclease sites;
restriction enzymes that recognize sites of relatively high dA/dT
composition cleave chloroplast DNA into many fragments while restriction
enzymes that recognize sites of relatively high dG/dC composition produce
fewer fragments. This observation is consistent with the low dG/dC content
of chloroplast DNA; low overall dG/dC content is reflected in the low
number of dG/dC rich restriction sites.

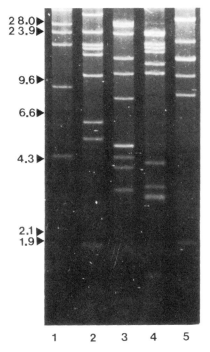

Fig. 1 - Gel electrophoretic patterns of *Vicia faba* chloroplast DNA
digested with restriction endonucleases. Restriction
patterns are shown for *Sma* I (lane 1), *Pst* I (lane 2),
Kpn I (lane 3), *Xho* I (lane 4) and *Sal* I (lane 5). The
size markers on the left of the diagram are kilobase
pairs. The data are summarized in Table I.

Gel electrophoretic patterns of *V. faba* chloroplast DNA digested
with a variety of restriction endonucleases are shown in Fig. 1.   A given
restriction endonuclease pattern also reflects genome size.   The molecular
weights of the individual DNA fragments can be added up to give the total
molecular weight of the *V. faba* chloroplast genome.   The restriction
fragment sizes produced by *Kpn* I, *Sma* I, *Pst* I, *Xho* I and *Sal* I are
summarized in Table I.   From this table the average overall genomic size
is estimated to be 123,000 base pairs (Ko *et al.,* 1983);   this size estimate
is in good agreement with the estimate from electron microscopic studies
(Koller and Delius, 1980).

Table I   Restriction enzyme analysis of chloroplast DNA

Pst I	*Xho* I	*Kpn* I	*Sal* I	*Sma* I
35.0 (P1)	20.5 (X1)	24.5 (K1)	31.0 (S1a&b)	50.0 (Sm1)
20.0 (P2)	19.5 (X2)	23.5 (K2)	16.5 (S2)	24.0 (Sm2)
15.5 (P3)	16.0 (K3)	20.0 (K3)	12.5 (S3a&b)	20.0 (Sm3)
14.5 (P4)	15.0 (X4)	12.5 (K4)	10.0 (S4)	16.0 (Sm4)
13.0 (P5)	14.0 (X5)	10.5 (K5)	8.5 (S5)	8.0 (Sm5)
10.0 (P6)	12.0 (X6)	8.1 (K6)	1.6 (S6)	4.4 (Sm6)
6.3 (P7)	10.5 (X7)	5.0 (K7a&b)		
5.2 (P8)	4.2 (X8)	4.3 (K8)		
1.8 (P9)	3.2 (x9)	4.1 (K9)		
1.1 (P10)	3.0 (X10a&b)	3.0 (K10)		
0.6 (P11)	1.3 (X11)	1.1 (K11)		
	0.6 (X12)	0.6 (K12)		
123,000	122,700	122,200	123,800	122,400

Native chloroplast DNA was digested with the restriction endonucleases
indicated.   Fragment sizes were determined using *Hind* III digested lambda
DNA as a size marker.   The fragments produced are given in kilobase pairs.
The totals are recorded in base pairs.   The letter-number codes in the
brackets indicate the fragment designations (Ko *et al.,* 1983; 1984).

Chloroplast genomes

Restriction endonuclease analysis, since it produces a species-specific pattern, can be used to examine genetic differences in cultivars or closely related species. Such techniques have been employed to study *Brassica, Nicotiana, Oenothera, Pisum* and other genera (Erickson, Straus and Beversdorf, 1983; Rhodes, Zhu and Kung, 1981; Gordon, Grouse, Bohnert and Herrmann, 1982; Palmer, Jorgeson and Thompson, 1985). Unfortunately, to our knowledge, such studies have not been applied to *V. faba*.

Physical Map of the Chloroplast Genome

Physical maps are necessary to provide reference points for the analysis of the chloroplast DNA; restriction endonucleases which recognize short DNA sequences and cleave at these sequences, provide convenient and reproducible reference points. The order that these restriction enzyme sites have on the circular chloroplast DNA can be determined by a variety of strategies (see Edelman, Hallick and Chua, 1982 for various ways). The physical map of the *Vicia faba* chloroplast genome was constructed using renaturation studies (Koller and Delius, 1980) and Southern blot hybridization (Ko *et al.*, 1983). The renaturation technique orders the restriction fragments by examining the relatedness of renatured fragments generated from two different restriction endonucleases (in this case *Kpn* I and *Sal* I). For the other technique, radiolabelled cloned *Pst* I fragments were used as a source of specific fragments to hydrize to restriction fragments generated by *Kpn* I, *Xho* I and *Sal* I. These fragments were separated on an agarose gel and transferred onto a nitrocellulose filter such as described by Southern (1975) prior to hybridization with radioactively labelled cloned *Pst I* fragments. Further details of the mapping schemes can be found in the cited reference. A physical map of the *Vicia faba* chloroplast genome is shown in Fig. 2.

GENETIC ORGANIZATION OF THE CHLOROPLAST GENOME

The *Vicia faba* chloroplast genome codes for one set of ribosomal RNAs and at least 31 tRNA species (Koller and Delius, 1980; Ko *et al.*, 1983; Mubumbila, Crouse and Weil, 1984). These genes occupy approximately 10% of the coding capacity of the chloroplast genome. The remaining sequence

124

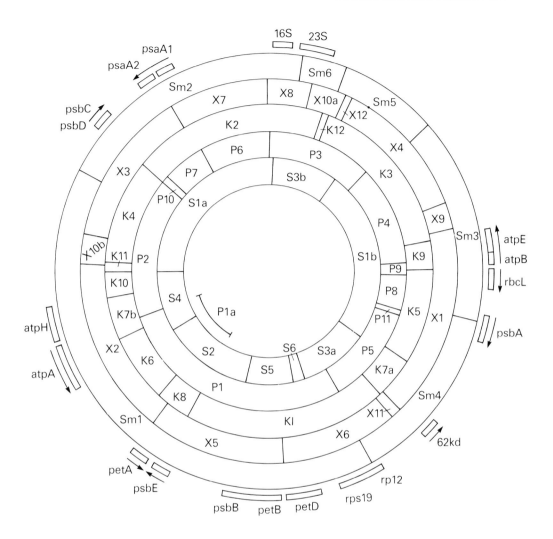

Fig. 2 - Physical map of the *Vicia faba* chloroplast genome. The
map shows restriction sites, starting from the inside the
map, for *Sal* I (designated 'S' fragments), *Pst* I
(designated 'P' fragments), *Kpn* I (designated 'K' fragments),
Xho I (designated 'X' fragments) and *Sma* I (designated
'Sm' fragments). The location and orientation of genes
are indicated. The fragment marked 'Pla' is one of the
fragments cloned to complete the clone bank (see Ko
et al., 1983). The map of *Sal* I sites was determined by
Koller and Delius, 1980.

have the capacity to code for 70 - 80 polypeptides, using an average polypeptide size of 40 kilodalton (equivalent to a gene of about 1,200 - 1,500 base pairs) per gene.

The most extensively mapped chloroplast genome is that of *Spinacia oleracea*. In addition to the rRNA and tRNA genes, approximately nineteen polypeptide genes have been mapped on the *Spinacia oleracea* chloroplast genome (Herrmann, Westhoff, Alt, Tittgen and Nelson, 1984). A set of eighteen chloroplast polypeptide genes have been mapped on the *Vicia faba* chloroplast genome; these are shown in Fig. 2 (Ko *et al.*, 1983; 1984; unpublished data).

Ribosomal RNA Genes
Chloroplasts are semi-autonomous organelles which possess their own protein synthesizing machinery. Translation in chloroplasts is mediated by ribosomes of the 70S type (Littleton, 1962) which are distinct from the 80S ribosomes of the cytoplasm. The 70S ribosomes contain four different RNA molecules. The 30S small subunit contains the 16S rRNA and the 50S large subunit contains the 23S, 5S and 4.5S rRNAs. Purified individual rRNAs, after radiolabelling, can be hybridized to chloroplast DNA fragments to determine the location of these genes on the chloroplast DNA physical map. It has been shown that in higher plants where physical maps are available, the rRNA genes are located close together and constitute one operon (Kossel, Natt, Strittmatter, Fritzsche, Gazdzicka-Jozefiak and Przybl, 1984). The genes for rRNAs are encoded within the inverted repeat sequences; therefore, two copies of these genes are present in most higher plant chloroplast chromosomes. However, since the *Vicia faba* chloroplast genome lacks the inverted repeat organization, only one set of rRNA genes are present per chloroplast DNA. The presence of only one set of ribosomal RNA genes was confirmed by R-loop analysis (Koller and Delius, 1980) and blot hybridization using *E. coli* 16S and 23S ribosomal RNAs as probes (Ko *et al.*, 1983).

The *Vicia faba* ribosomal RNAs are localised close together in *Sal* I fragment S3a and *Pst* I fragment P3 (Koller and Delius, 1980; Ko *et al.*, 1983). The arrangement of the ribosomal RNAs is similar to bacteria, such as *E. coli*, and the blue-green alga, *Anacystis nidulans* (Tomioka *et al.*, 1981; Kossel *et al.*, 1984). The order is as follows: 16S rDNA - spacer -

23S rDNA - spacer - 4.5S - (spacer) - 5S rDNA. The 16S rDNA (1580 base pairs long) is located 1609 bp from one end of the S3a fragment. The spacer between 16S and 23S rDNA is 2372 base pairs long; the ribosomal spacer of most higher plant chloroplast genomes, is greater than 2000 base pairs. The 23S rRNA gene (3162 base pairs long) is 3247 base pairs from the other end of S3a DNA fragment. These measurements were determined by electron microscopy and R-loop analysis (Koller and Delius, 1980). The length of the spacer between 23S and 4.5/5S rDNA has not been determined.

The ribosomal RNA genes have been cloned (Ko *et al.*, 1983; Endo, Kusuda and Sugiura, 1982) in the plasmid vectors pBR322 and pBR325. None of the rRNA genes have been sequenced or studied in detail so far; however, studies in other chloroplast genomes have shown that chloroplast rRNA genes have a high degree of homology with *E. coli* rRNA sequences. The 16S rRNA is 72% - 74% homologous and the 23S rRNA is only slightly less (67% for tobacco, 71% for maize) homologous with the *E. coli* ribosomal RNA genes (Schwarz and Kossel, 1980; Edwards and Kossel, 1981; Takaiwa and Sugiura, 1982; Tohdoh and Sugiura, 1982).

Transfer RNA Genes

Transfer RNAs coded by the chloroplast genome are distinct from tRNAs found in the cytoplasm and mitochondria. Unlike mitochondria which can deviate from the "universal code" (Bibb *et al.*, 1981), chloroplast tRNA adhere strictly to universal codon usage for translation. Unlike the cytoplasm, the chloroplasts possess a relatively small set of tRNAs. The number of tRNA genes is approximately the minimum number expected, according to the wobble hypothesis (Crick, 1966), to read all sixty-one sense codons and support translation. Individual chloroplast tRNAs specific for all 20 amino acids have not yet been isolated or characterized, however, there is evidence that a complete set of tRNA genes corresponding to all 20 amino acids is present (Meeker and Tewari, 1980). A minimum of thirty one tRNA species from *Vicia faba* chloroplasts has been fractionated by two dimensional polyacrylamide gel electrophoresis. Twenty five of these tRNAs were identified by aminoacylation and found to be specific for 17 amino acids. Six tRNAs are still unidentified and may be specific for the three remaining amino acids (cysteine, glutamic acid and glutamine).

A minimum of 30 tRNA genes and some unidentified tRNAs have been located on the physical map of *Vicia faba* chloroplast DNA. This was achieved by combining results of homologous and heterologous hybridization using both individual *Vicia faba* and *Phaseolus vulgaris* tRNAs (Mubumbila *et al.,* 1984). In most experiments, the individual tRNAs were radiolabelled by *in vitro* procedures and hybridized to chloroplast DNA fragments generated by the restriction enzymes, *Sal* I and *Kpn* I. The hybridization results show that regions containing tRNA genes are distributed around most of the circular chloroplast genome (Mubumbila *et al.,* 1984).

In order to study the structure and other features of tRNA genes, it is necessary to determine the nucleotide sequence. Sequencing data indicate that chloroplast tRNA genes possess both prokaryotic and eukaroytic features. The most prominent prokaryotic feature is the relatively high sequence homology (70%) with corresponding bacterial tRNA genes. Other prokaryotic features include the ability to charge chloroplast tRNAs with bacterial amino-acyl tRNA synthetases and the presence of sequences in the 5' flanking region resembling the "-35" and "-lO' regions of bacterial promoters. These upstream regions probably also act as promoters for chloroplast tRNA genes. In contrast to these prokaryotic features, a number of tRNA genes are split by large introns (450-lOOO base pairs), a feature observed in nuclear tRNA genes from eukaryotes. The present data for *Vicia faba* chloroplast tRNA genes do not appear to deviate from these findings. A set of three tRNA genes specific for glutamine (Glu), tyrosine (Tyr) and threonine (Thr) have been sequenced (Kuntz, Weil and Steinmetz, 1984). All three genes are found within a 2014 base pair *Bam* HI fragment. This *Bam* HI fragment (called '*Bam* 19') is located within the overlapping *Sla#K4* region. The tRNA^{Glu} (75 base pairs) and tRNA^{Tyr} (87 base pairs) genes are separated by 6O base pairs and may appear to be part of the same transcriptional unit. the tRNA^{Thr} (75 base pairs) gene is located on the opposite strand, 876 base pairs away from the tRNA^{Glu} gene. The finding of a tRNA^{Glu} gene was unexpected since it had never been mapped on any of the higher plant chloroplast genomes which have been studied; technical problems encountered with the identification of the purified tRNAs have so far prevented the mapping of this gene.

Interesting features have been revealed in nucleotide sequencing studies of two tRNA^{Leu} genes (Bonnard, Michel, Weil and Steinmetz, 1984). The gene encoding the tRNA$^{Leu}_{UAA}$ has been located on a 5.1 kbp *Bam* HI fragment. Unlike the tRNA genes so far sequenced for *V. faba,* this gene is 536 base pairs long and the anticodon region is interrupted by a four hundred and fifty one base pair long intron. The intron shows high sequence homology over about 100 base pairs from each end with the corresponding regions of the maize chloroplast tRNA$_{UAA}^{Leu}$ intron. It has been suggested that these conserved sequences are involved in the splicing reactions necessary to produce the final tRNA molecule. The tRNA$_{CAA}^{Leu}$ gene is located in the same *Bam* HI fragment, 443 base pairs away and is transcribed in the opposite direction from the complementary strand. The intergenic region, between the two tRNA^{Leu} genes, contains a number of short repeats and a partial copy of the intron-split tRNA$_{UAA}^{Leu}$ gene. The partial copy includes 100 bp of the 5' flanking region, 35bp of the 5' exon and the first 42 bp of the intron. It has been suggested that duplication occurred upon the rearrangement of the two tRNA^{Leu} genes in *Vicia faba* or upon the deletion of the inverted repeat sequences. In all other higher plant chloroplast genomes with inverted repeats, the two tRNA^{Leu} genes are separable by a large amount of DNA. One is located in the inverted repeat region, and the other is in the large single copy region; the gene for tRNA^{Phe} is encoded by the same strand as tRNA$_{UAA}^{Leu}$ 100 bp downstream and the direction of transcription is the same as tRNA$_{UAA}^{Leu}$. The nucleotide sequence of the tRNA^{Phe} has also been determined in *V. faba*.

Chloroplast Polypeptide Genes

In *Vicia faba,* a total of eighteen chloroplast polypeptide genes have been located on the chloroplast DNA. Their locations have been determined by using gene specific probes constructed from heterologous sources, such as spinach, pea, mung bean and tobacco chloroplast DNA (Ko *et al.*, 1984, Shinozaki, Sun and Sugiura, 1984, Ko *et al.*, 1983). Studies using gene probes from unrelated plant chloroplast DNA are feasible due to the high degree of sequence conservation for plastid genes from different species. The gene probes were radiolabelled and hybridized to Southern transfers of chloroplast DNA fragments generated by *Kpn* I, *Pst* I, *Sal* I and *Xho* I. Genes have been mapped for subunits involved in the enzyme ribulose-1, 5-bis-phosphate carboxylase, the cytochrome b$_6$/f complex, ATP synthase, ribosomes, photosystems I and II (Ko *et al.*, manuscript in preparation; Ko *et al.*, 1984; Ko *et al.*, 1983). The directions of transcription for eleven of the genes have been determined by using gene probes from the 5' and 3' ends of the genes in question. The position and direction of transcriptions are indicated on the physical map (Fig. 2).

Complete nucleotide sequences of identified polypeptide genes encoded by the *V. faba* chloroplast genome have not been reported so far. The nucleotide sequence of the spacer region between the genes for the large subunit of ribulose-1,5-bishosphate carboxylase/oxygenase and the B subunit of ATP synthase has been reported by Shinozaki *et al.*, (1984). The spacer region is 795 base pairs long. Prokaryotic signal sequences corresponding to the "Pribnow's box" and the "-35 region" of *E. coli* genes were found upstream from both gene coding regions. This observation once again indicates that transcriptional initiation of chloroplast genes resembles that of prokaryotic genes.

Most of the chloroplast encoded polypeptides that have been mapped on the chloroplast genome to date are polypeptides of known function that have been isolated and are sufficiently antigenic for the formation antibodies. These antibodies have been used to identify which cloned chloroplast DNA fragment is able to produce the polypeptide in question when tested in *E. coli in vitro* expression systems (Edelman, Hallick and Chua, 1982; Herrmann *et al.*, 1984). As indicated earlier the chloroplast genome has a greater coding capacity than that required to code for the polypeptides that have already been mapped.

Studies to locate the unidentified genes involve one of two strategies. First, one can determine which areas of the chloroplast are transcribed by Northern blot hybridization studies; in these studies radioactively labelled cloned chloroplast DNA fragments are hybridized to nitrocellulose immobilized plant RNA. Second, one can assay the number and molecular weight of polypeptides that are produced *in vitro* from every clone of a complete chloroplast DNA library. A complete clone bank of the *Vicia faba* chloroplast DNA genome has been established (Ko *et al.*, 1983) to provide a source of DNA for these studies. Fragments generated by *Pst* I, *Kpn* I and *Sal* I have been inserted into the plasmid vectors pBR322 (Bolivar *et al.*, 1977) and pDPL13 (Gendel *et al.*, 1985). Any of these cloned regions can be further dissected into smaller fragments by subcloning for more detailed studies.

Cloned fragments representing the entire *Vicia faba* chloroplast genome have been tested in an *E. coli in vitro* transcription-translation system. A large number of polypeptides were expressed (Ko *et al.*, manuscript in preparation). The distribution of expression products was not even along the chloroplast genome. The regions represented by P1, P2 and P4 were very active in expressing products. Low levels of expression were found in regions represented by P3, P6 and P7. Although these results do not reveal the identity and function of the expressed polypeptides, they supply information on potential gene coding sequences and the size of the gene product. The potential gene coding sequences can then be studied by other strategies.

One of these potential coding sequences is currently under investigation because the *V. faba* gene(s) in question apparently originated from the inverted-repeated sequences. Hybridization studies indicate that this expressed region is localized to the middle of the inverted-repeats of *Brassica napus* (Ko *et al.*, unpublished results). This coding region was first detected by testing the cloned fragment P5, S3a and K7a in the *E. coli in vitro* and maxicell expression system. Polypeptides with M_R values of 62,000 and 43,000 were expressed. The region containing these potential coding sequences has been located within a 2000 bp *Eco*RI fragment and if both products represent separate chloroplast polypeptides their genetic sequences must be extensively overlapped. The direction of transcription (as shown on the map) has been determined by testing various subclones of the 2000 base pair *Eco*RI fragment for *in vitro* expression. The nucleotide sequence of the entire *Eco*RI fragment has been determined; however, the identity and function of the polypeptides remain to be elucidated.

Higher plant chloroplast polypeptide genes are generally uninterrupted by intervening sequences except in two reported cases. In one case, a 666 base pair insertion was found in the ribosomal protein gene, *rpl2*, of *Nicotiana debneyi* meanwhile the same gene sequence in *Spinacia oleracea* was un-interrupted (Zurawski *et al.*, 1984). In a more recent case, the gene coding for the CF_0 subunit I of ATP synthase from wheat contains a 823 base pair intron (Bird, Koller, Aufret, Hutley, Howe, Dyer and Gray, 1985). There is also the possibility of an intron in the CF_0 subunit I gene of *Spinacia oleracea* but the results have not been conclusive (Westhoff, Alt, Nelson and Herrmann, 1985). A systematic search for introns in the *Vicia faba* chloroplast genome has been reported by Koller and Delius in 1984. Circular chloroplast DNA was hybridized with homologous RNA. Results show that almost all hybrids found were uninterrupted. Two transcripts with two introns each and two others with one intron each were identified. Two spliced transcripts were located in the middle of K2, about 2000 base pairs upstream of the rRNA operon. One hybrid is about 1,200 base pairs long and has one intron, the other is about 5,800 base pairs long and has two introns. The other transcript with two introns was located in the K1 fragment. All of the introns of these three transcripts have the same size of about 800 base pairs. A sixth intron of about 600 base pairs was found in fragment S2. No hybrids with introns were found in the other fragments. All spliced transcripts found in the *Vicia faba* chloroplasts are longer than 1,000 base pairs; however, it has not been ruled out that these introns are part of a spliced tRNA gene (see the section on tRNA) that is co-transcribed and close proximity to a protein gene. It, therefore, remains to be seen whether the long transcripts that form intron loops with *Vicia faba* chloroplast DNA are really functional mRNAs of chloroplast polypeptide genes.

EVOLUTIONARY ASPECTS OF THE CHLOROPLAST GENOME

As a general rule the genetic arrangement of the chloroplast genome has been strongly conserved within the vascular plants. Few structural alterations have been reported from ferns to angiosperms. The greatest single rearrangement in chloroplast genomic evolution is the deletion of one of the inverted repeated segments in a group of legumes which includes *Vicia faba*. Surprisingly, this group of plants also appears to have undergone many other rearrangements as well.

The simplest change from the conserved chloroplast genome with the inverted repeated structure to one lacking the inverted repeat is seen in alfalfa (*Medicago sativa*). The alfalfa chloroplast genome differs from chloroplast genomes with inverted repeats by the simple deletion of one entire segment of the inverted repeat. Studies on the arrangement of restriction fragments in *Vicia faba* indicate that *V. faba* may have been derived from an alfalfa-like ancestral chloroplast genome by two large overlapping inversions (Palmer, unpublished data). Other legumes of this group have undergone even more extensive rearrangements; for example, the chloroplast DNA of *Pisum sativum* differs from that of alfalfa by approximately 12 inversions and the chloroplast genome of clover has sustained an undetermined number of complex rearrangements (Palmer, unpublished data).

Rearrangement of the Surviving Inverted Repeat Region

Studies on the rearrangement of chloroplast DNA of *Vicia faba* (Palmer and Thompson, 1982; Shinozaki *et al.*, 1984), have shown that the region corresponding to the "inverted repeat DNA" is separated into two different regions of the chloroplast genome of *V. faba*. The inverted repeat-large single copy junction is located in a more or less fixed position within the *rps 19* gene, a small subunit ribosomal polypeptide gene (Palmer, 1985); radioactive *rps 19* gene probe from spinach hybridizes to the Pl/S3a/X6 region in *Vicia faba* chloroplast DNA. We have shown that a polypeptide gene that is located in the middle of the inverted repeat sequence, is found as one copy within the K7a fragment of *V. faba*. The other end of this run of surviving "inverted-repeats" is probably marked by the position of the tRNA$_{CAA}^{Leu}$ gene. As discussed in the section on tRNA genes this gene is normally found in the middle region of "inverted-repeats"; in *V. faba* this gene is located quite close to the tRNA$_{UAA}^{Leu}$ and tRNA^{Phe} genes that are usually located well within the large single copy region of the ancestral chloroplast genome. The rest of the inverted repeat region, that contains the rRNA cistrons, is contained within the P3/P6 region (Palmer and Thompson, 1982; Ko *et. al.*, 1983; Shinozaki *et al.*, 1984). Shinozaki *et al.*, (1984) indicated a possible *rps 19* gene in the P6 fragment; however, in the absence of some as yet unknown duplication event, these results conflict with the information presented in the physical map (Fig. 2). The P6 region deserves further study to define the rearrangement of the inverted repeat region.

To date no other segments corresponding to the inverted repeated region have been found in *V. faba*. The possibility exists, however, that small regions have escaped detection; Mubumbila *et al.*, (1984) have located a tRNA^{Leu} gene in the KlO/Kll region of *V. faba* using isolated tRNA's but the actual sequences involved in the reaction have yet to be determined. Until there is evidence to the contrary, it would appear then that the DNA sequences corresponding to one of the inverted repeated regions are divided between the P3/P6 and the S3a regions only.

Rearrangement of the Small Single Copy Region

The region corresponding to the small single copy sequence of the ancestral type of chloroplast genome, has also been rearranged into at least two distinct areas of the *Vicia faba* chloroplast genome. Probes constructed

from the small single copy region of *Nicotiana tabacum* and *Vigna radiata*
were used to determine the corresponding regions in *Vicia faba* chloroplast
DNA (Palmer and Thompson, 1982; Shinozaki *et al.*, 1984). The homologous
regions were located in the K7a/K5 and K3 fragments. Further studies
have recently been carried out using probes constructed from *Bam* HI
fragments from the small single copy region of *Brassica napus* (Ko *et al.*,
unpublished observations). The results generally confirmed those reported
by Palmer and Thompson (1982) and Shinozaki *et al.*, (1984); small single
copy sequences are found in the overlapping P5/K5/K7a area and in the
P4/K3 region. However, an unexpected result was revealed by the more
detailed study; at least one third of the ancestral small single copy region
is absent from the *Vicia faba* chloroplast genome. Studies are currently
underway to determine how extensive the loss is and which genes may have been
lost. To date no genes have been identified in the small single copy region.
The identification of genes would surely aid in the rearrangement studies.

Rearrangement of the Large Single Copy Region

The region corresponding to the small single copy sequence of the ancestral
type of chloroplast genome, has also been rearranged into at least two
distinct areas of the *Vicia faba* chloroplast genome. Probes constructed
from the small single copy region of *Nicotiana tabacum* and *Vigna radiata*
were used to determine the corresponding regions in *Vicia faba* chloroplast
DNA (Palmer and Thompson, 1982; Shinozaki *et al.*, 1984). The homologous
regions were located in the K7a/K5 and K3 fragments. Further studies have
recently been carried out using probes constructed from *Bam* HI fragments
from the small single copy region of *Brassica napus* (Ko *et al.*, unpublished
observations). The results generally confirmed those reported by Palmer
and Thompson (1982) and Shinozaki *et al.*, (1984); small single copy sequences
are found in the overlapping P5/K5/K7a area and in the P4/K3 region.
However, an unexpected result was revealed by the more detailed study;
at least one third of the ancestral small single copy region is absent from
the *Vicia faba* chloroplast genome. Studies are currently underway to
determine how extensive the loss is and which genes may have been lost. To
date no genes have been identified in the small single copy region. The
identification of genes would surely aid in the rearrangement studies.

Rearrangement of the Large Single Copy Region

Because the large single-copy region contains most of the identified chloroplast genes that code for polypeptides, this section will focus on the rearrangement of gene clusters in the *Vicia faba* chloroplast DNA relative to the ancestral type of chloroplast genome represented by *Spinacea oleracea*. Chloroplast gene arrangement will also be compared between *Vicia faba* and the other well studied legume chloroplast genome, *Pisum sativum*. The differences in genetic arrangement will be discussed in terms of overall organisation and the significance of gene clusters/blocks. Despite the loss of the inverted repeat and some of the small single copy region in *Vicia faba* and perhaps in *Pisum sativum* as well, all the genes identified so far in *Spinacia oleracea* have also been found in *Vicia faba* and *Pisum sativum*.

The gene order representing *Spinacia oleracea* is highly conserved among most higher plant chloroplast DNAs. "Global" cross-hybridization studies and gene mapping studies (Palmer, 1985) were used to assay the extent of positional conservation among different chloroplast DNAs. Of the 30 families of angiosperms examined, 24 appear to have the same gene order as spinach. Gene order differences in the altered genomes can, in several cases, be explained in terms of one or two simple inversional switches. The explanation for the rearrangement of genes in *Vicia faba* and especially *Pisum sativum* is more complicated.

"Global" hybridization studies revealed that *Vicia faba* chloroplast DNA probably has evolved from an alfalfa-like ancestral genome by two inversions. Like *Vicia faba* alfalfa shares the loss of the inverted repeat structure, which is otherwise universally present among higher plants such as spinach. However, unlike *Vicia faba*, all the genes of alfalfa are co-linear when compared to legumes such as mung bean, common bean and soya bean, that have not lost one of the inverted-repeats. The gene order in mung bean chloroplast DNA differs from spinach by an 50 kilobase pair inversion. This inversion in mung bean reverses the gene order between the *psbA* and *petA* genes and places the *rbcL-atpB,E* genes next to the *psbA* gene and the *atpH,atpA* genes next to the *petA* gene. In *Vicia faba*, two further inversions create more gene order changes. One inversion appears to involve a fragment from the 5' end of the 16S rRNA gene to the 5' of *atpE* gene. This inversion changes the gene order with respect to the 16S-23S rRNA genes; genes initially located on one side of the ribosomal RNA genes are on the other side. A smaller inversion within this fragment reverses the *psbA-rbcL-atpB,E* gene order with respect to the ribosomal RNA genes. The *atpB,E* genes are close to the 23S rRNA instead of the *psbA* gene. The smaller inversion also resulted in the separation of the once co-linear small single copy region.

The rearrangement of genes in *Vicia faba* chloroplast DNA is relatively simple when compared to the rearrangements found in *Pisum sativum* chloroplast DNA. The genetic order in pea may have resulted from a minimum of 12 inversions from an alfalfa-like ancestral genome. Despite the extensive rearrangement in broad bean and pea, certain gene clusters/blocks remain intact blocks like *atpB,E-rbcL,* or *petD-petB-psbB* remain together. The conservation of gene clusters suggest that the genes in question may be under a common controlling element and transcribed into polycistronic messages. Indeed Northern blot hybridization studies for all cloned fragments of a complete *V. faba* chloroplast DNA bank indicate that large molecular weight mRNAs occur commonly in the chloroplast of *V. faba* and thus many genes may be contranscribed to form polycistronic RNA.

CONCLUDING REMARKS

The chloroplast genome of most vascular plants contains inverted repeated sequences separated by single copy sequences. *V. faba* belongs to a special group of plants whose chloroplast genome is missing one of the inverted-repeated regions. Chromosomal mapping indicates that in addition to this deletion, the chloroplast genome of *V. faba* has probably sustained only two other rearrangements during its evolution. These changes result from two inversions that break up the ancestral inverted-repeated region and alter the location of polypeptide encoding genes relative to the ribosomal cistron. Because the deletion event involved part of the small single copy region, studies of chloroplast DNA from *V. faba* have identified areas of the ancestral chloroplast genome that merit special study to determine whether or not these sequences have been lost to *V. faba* or simply transferred to the nucleus like other ancestral chloroplast genes, ie. the small subunit of ribulose-1,5-bisphosphate carboxylase. If the sequences have indeed been lost during evolution, future studies on the function of these sequences should help to elucidate other as yet unknown differences between chloroplasts that contain inverted repeated sequences and those that do not.

Vicia faba DNA also provides an opportunity to study sequences that border the sites of chromosomal rearrangement. At least one of these rearrangements involves the central area of the ancestral inverted-repeated sequences. This has real significance to our understanding of ancestral chloroplast stability and of chloroplast site specific recombination because site specific recombination appears to play an important role in normal chloroplasts. It is now generally assumed that recombination between specific sites in the inverted-repeats is responsible for the two forms of the chloroplast genome that have been demonstrated in the cyanelle of *Cyanophora* and of the common bean, *Phaseolus vulgaris* (Bohnert and Loffelhardt, 1982; Palmer, 1983). These two forms differ by the orientations of genes found in the two single copy regions; in one form the entire small single copy region has the reverse order to the other form. Palmer (1985) has postulated that the recombination sites of the inverted-repeat and other secondary recombination sites throughout the genome are responsible for the large number of chloroplast genomic rearrangements found in legumes. *V. faba* provides the ideal system to study such sites because there are few well defined rearrangements.

Finally, the chloroplast genome of *V. faba* has been studied in greater detail than any other legume lacking one of the inverted repeats. The information gathered to date provides a strong basis for continued studies on chloroplast molecular genetics. Perhaps the *V. faba* chloroplast will be the system in which we will discover the functional significance of inverted repeats in the ancestral genomes.

ACKNOWLEDGEMENTS

The authors would like to thank Dr. J.D. Palmer for making unpublished data available.

REFERENCES

Bibby, M.J., Van Etten, R.A., Wright, C.T., Walberg, M.W., and Clayton, D.A. (1981). Sequence and gene organisation of mouse mitochondrial DNA *Cell* 26:167 -180.

Bird, C.R., Koller, B., Auffret, A.D., Huttly, A.K., Howe, C.J., Dyer, T.A., and Gray, J.C. (1985). The wheat chloroplast gene for CF_o subunit I of ATP synthase contains large intron. *EMBO 4*: 1381 - 1388.

Bohnert, H.J., Crouse, E.J., and Schmitt, J.M. (1982). Organisation and expression of plastid genomes. *Encycl. Plant Physiol. 14B*: 475 - 530.

Bohnert, H.J., and Loffelhardt, W. (1982). Cyanelle DNA from *Cyanophora paradoxa* exists in two forms due to the intramolecular recombination. *FEBS Lett. 150*: 403 - 406.

Bolivar, F., Rodriguez, R.L., Greene, P.J., Betlach, M.C., Heynecker, H.W., Boyer, H.W., Crosa, J.H., and Falkow, S. (1977). Construction and characterisation of a cloning vehicles II. A multipurpose cloning system. *Gene 2*: 95 - 113.

Bonnard, G., Michel, F., Weil, J.H., and Steimetz, A. (1984). Nucleotide sequence of the split $tRNA^{Leu}_{UAA}$ gene from *Vicia faba* chloroplasts: evidence for structural homologies of the chloroplast tRNA intron with the intron from the autosplicable *Tetrahymena* ribosomal RNA precursor. *Mol. Gen. Genet. 194*: 330 - 336.

Bottomley, W., and Bohnert, H.J. (1982). The biosynthesis of chloroplast proteins. *Encycl. Plant Physiol. 14B*: 531 - 596.

Crick, F.H.C. (1966). Codon - anticodon pairing: the wobble hyothesis. *J. Mol. Biol. 19*: 548 - 555.

Edelman, M., Hallick, R., and Chua, N.H. (eds.) (1982). *Methods in chloroplast molecular biology*. Elsevier, Amsterdam.

Edwards, K., and Kossel, H. (1981). The rRNA operon from *Zea mays* chloroplasts: nucleotide sequence of 23S rDNA and its homology with *E. coli* 23S rDNA. *Nucl. Acids Res. 9*, 253 - 2869.

Ellis, R.J. (1981). Chloroplast proteins synthesis, transport and assembly. *Ann. Rev. Plant Physiol. 32*: 111 - 137.

Erickson, L.R., Straus, N.A., and Bevesdorf, W.D. (1983). Restriction patterns reveal origins of chloroplast in *Brassica* amphiploids. *Theor. Appl. Genet. 65*: 201 - 206.

Gendel, S.M.G., Straus, N., Pulleyblank, D., and Williams, J.P. (1983). A novel shuttle cloning vector for the cyanobacterium *Anacystis nidulans*. *FEMS Microbiol. Lett. 19*: 291 - 294.

Gordon, K.H.J., Crouse, E.J., Bohnert, H.J., and Hermann, R.G. (1982). Physical mapping of difference in chloroplast DNA of the five wild-type plastomes in *Oenothera* Subsection *Euonethera. Theor. Appl. Genet. 61*: 373 - 384.

Groot, G.S.P. (1984). Molecular form and function of chloroplast DNA in higher plants. In *Molecular form and function of the plant genome.* (eds. van Vloten-Doting, L., Groot, G.S.P., Hall, T.C. pp 175 - 181). Plenum Press, Amsterdam.

Herrmann, R.G., Westhoff, P., Alt, J., Tittgen, J., and Nelson, N. (1984). Thylakoid membrane proteins and their genes. In *Molecular form and function of the plant genome.* (eds. van Vloten-Doting, L., Groot, G.S.P., Hall, T.C. pp 233 - 256). Plenum Press, Amsterdam.

Ko, K., Straus, N.A., and Williams, J.P. (1984). The localisation and orientation of specific genes in the chloroplast chromosome of *Vicia faba. Curr. Genet. 8*: 359 - 367.

Ko, K., Strauss, N.A., and Williams, J.P. (1983). Mapping the chloroplast DNA of *Vicia faba . Curr. Genet. 7*: 255 - 263.

Koller, B., and Delius, H. (1984). Intervening sequences in chloroplast genomes. *Cell 29*: 613 - 622.

Koller, B., and Delius, H. (1980). *Vicia faba* chloroplast DNA has only one set of ribosomal RNA genes as shown by partial denaturation mapping and R-looping analysis. *Mol. Gen. Genet. 178*: 261 - 269.

Kossel, K., Natt, E., Strittmatter, G., Fritzsche, E., Gozdzicka-Josefiak, A., and Przybyl, D. (1984). In *Molecular form and function of the plant genome.* (eds. van Vloting-Doting, L., Groot, G.S.P., Hall, T.C. pp 183 -

Kung, S.D., and Williams, J.P. (1969). Chloroplast DNA from broad bean. 197. *Biochim. Biophys. Acta 195*: 433 - 445.

Kuntz, N., Weil, J.H., and Steinmetz, A. (1984). Nucleotide sequence of a 2kbp *Bam* HI fragment of *Vicia faba* chloroplast DNA containing the genes for threonine, glutamine acid and tyrosine transfer tRNA. *Nucl. Acids Res. 12* 5037 - 5047.

Lyttleton, J.W. (1962). Isolation of ribosomes from spinach chloroplasts. *Exp. Cell Res. 26*: 312 - 317.

Mandel, M., Schildkraut, C.L., and Marmur, J. (1968). Use of CsCl density gradient analysis for determining the guanine plus cytosine content of DNA. *Methods Enzymol, 12b*: 184 - 195.

Meeker, R., and Tewari, K.K. (1980). Transfer ribonucleic acid genes in the chloroplast deoxyribonucleic acid of pea leaves. *Biochemistry 19*: 5973 - 5981.

Mubumbila, M., Crouse, E.J., and Weil, J.H. (1984). Transfer RNAs and tRNA genes of *Vicia faba* chloroplasts. *Curr. Genet.* 8: 379 - 385.

Palmer, J.D. (1985). Comparative organisation of chloroplast genomes. *Ann. Rev. Genet.* In press.

Palmer, J.D. (1983). Chloroplast DNA exists in two orientations. *Nature* 26: 312 - 317.

Palmer, J.D., and Thompson, W.F. (1982). Chloroplast DNA rearrangements are more frequent when a large inverted repeat sequence is lost. *Cell* 29: 537 - 550.

Palmer, J.D., Jorgensen, R.A., and Thompson, W.F. (1985). Chloroplast DNA variation and evolution in *Pisum*: Patterns of change and phylogenetic analysis. *Genetics 109*: 195 - 213.

Rhodes, P.R., Zhu, Y.S., and Kung, K.D. (1981). *Nicotiana* chlorolast genome I. Chloroplast DNA diversity. *Mol. Gen. Genet. 182*: 106 - 111.

Schwarz, Z., and Kossel, H. (1980). The primary structure of 16S rDNA from *Zea mays* chloroplast is homologous to *E.coli* 16S rRNA. *Nature 283*: 739 - 742.

Shinozaki, K., Sun, C.R., and Sugiura, M. (1984). Gene organisation of chloroplast DNA from the broad bean *Vicia faba*. *Mol. Gen. Genet. 19*: 363 - 367.

Southern, E.M. (1975). Detection of specific sequences among DNA fragments separated by gel electrophoresis. *J. Mol. Biol. 98*: 503 - 517.

Sun, C.R., Endo, T., Kusuda, M., and Sugiura, M. (1982). Molecular cloning of the genes for ribosomal DNAs from broad bean chloroplast DNA. *Jpn. J. Genet.* 57: 397 - 402.

Takaiwa, F., and Sugiura, M. (1982). The complete nucleotide sequence of a 23S ribosomal RNA gene from tobacco chloroplasts. *Eur. J. Biochem. 124*: 13 - 19.

Tohdoh, N., and Sugiura, M. (1982). The complete nucleotide sequence of a 16S ribosomal RNA gene from tobacco chloroplasts. *Gene 17*: 213 - 218.

Tomioka, N., Shinozaki, K., and Sugiura, M. (1981). Molecular cloning and characterisation of ribosomal RNA genes from blue-green alga *Anacystis nidulans*. *Mol. Gen. Genet. 184*: 359 - 363.

Westhoff, P., Alt, J., Nelson, N., and Herrmann, R.G. (1985). Genes and transcripts for the ATP synthase CF_o subunits I and II from spinach thylakoid membranes. *Mol. Gen. Genet. 199*: 290 - 299.

Whitfield, P.R., Bottomley, W. (1983). Organisation and structure of chloroplast genes. *Ann. Rev. Plant Physiol.* *34*: 219 - 310.

Zurawski, G., Bottomley, W., and Whitfeld, P.R. (1984). Junctions of the large single copy region and the inverted repeats in *Spinacia oleracea* and *Nicotiana debneyi* chloroplast DNA sequence of the genes for tRNA[His] and the ribosomal proteins S19 and L2. *Nucl. Acids Res..* *12*: 6547 - 6558.

6   The promoters and terminators of chloroplast genes

S. D. Kung, S. Akada, S. Mongkolsuk, X. F. Kong, C. M. Lin and P. S. Lovett

ABSTRACT

Chloroplast genomes are organized into many transcriptional units. Each unit
contains single or multiple genes and requires defined signals to initiate
transcription at one end and to terminate it at the other. Such signals are
known as promoters and terminators. The promoter is a region of DNA
involved in the binding of RNA polymerase to initiate transcription. It
consists of two regions of conserved sequences, located about 10 and 35 bp
upstream of the transcription start point and separated by an optimal distance
of 17 bp. The terminator is a region of DNA containing recognition signal(s)
for termination of transcription by RNA polymerase.

INTRODUCTION

Chloroplast genomes are organized into many transcriptional units. Each unit
contains single or multiple genes and requires defined signals to initiate
transcription at one end and to terminate it at the other. Such signals
are known as promoters and terminators, respectively (Rosenberg and Court,
1979). The promoter is defined as a region of DNA involved in the binding
of RNA polymerase to initiate transcription. The chloroplast promoters as
in prokaryotes are made up of two domains known as the -10 and -35 regions
separated by an optimal distance of 17bp (Rosenberg and Court, 1979;
Hawley and McClure, 1983). The terminator is a region of DNA containing
recognition signal(s) for termination of transcription by RNA polymerase.
There are two types of terminators, Rho-independent terminators function in
the absence of Rho, whereas the function of Rho-dependent terminator
requires Rho (Roberts 1969). Generally, the Rho-independent terminators
consist of the dyad symmetry capable of forming stem and loop structures
and a run of U residues (Platt, 1981).

Currently, our knowledge of chloroplast promoters is advancing rapidly. Promoters of many chloroplast genes have been structurally well defined and thoroughly analyzed (Kung and Lin, 1985). This was accomplished mainly by using a gene fusion technique (Kong, Levett and Kung, 1984; Rosenberg, Chepelinsky and McKenny, 1983), an *in vitro* homologous transcription system (Orozco, Mullet and Chua, 1985; Link, 1984), deletion and mutagenesis (Link, 1984). On the other hand, there is little information on the chloroplast terminators. Only very recently, the 3' flanking regions of selected chloroplast genes have been sequenced and contain structures resembling the prokaryotic termination sites (Whitfield and Bottomley, 1983; Zurawski, Perret, Bottomley and Whitfield, 1981; McIntosh, Poulsen and Bogorad, 1980; Shinozaki and Sugiura, 1981; Bovenberg, Koes, Kool and Nijkamp, 1984; Zurawski, Bohnert, Whitfield and Bottomley, 1982; Sugita and Sugiura, 1984; Link and Langridge, 1984). Whether such **structures** function as terminators in any system is an interesting question. This chapter provides a survey and a summary of the current information on promoters and terminators of chloroplast genes from higher plants.

PROMOTERS AND TERMINATORS

The transcriptional cycle consists of three main steps: initiation, elongation and termination. RNA chain initiation includes the binding of RNA polymerase to the promoter, formation of a polymerase-DNA complex and catalysis of the first 3'-5' internucleotide bond. RNA chain elongation requires translocation of RNA polymerase along the DNA template. Finally, RNA chain termination involves dissociation of the complex. Obviously, the initiation of transcription is a crucial stage at which gene expression can be regulated. However, the contribution of the terminators in regulating the gene expression should not be overlooked. The identification and analysis of attenuator sites preceeding the gene clusters has demonstrated that the extent of gene expression can also be modulated by the termination, independent of promoter/operator regulation (Yanofsky, 1981). Such modulation is achieved by controlling the levels of transcriptional read-through at terminators.

Recently, both promoters and terminators have been the subject of intensive investigations in many systems particularly in prokaryotes.

After analysing the DNA sequences of 112 well-defined promoters, a structure-function relationship of prokaryotic promoters (consensus sequences) was established (Rosenberg and Court, 1979; Hawley and McClure, 1983; Travers, 1984). There are two main domains of DNA sequences within the promoter which determine its overall effectiveness. One is the AT-rich sequence around the -35 region which has often been referred to as the recognition site and the other is the sequence around the -10 region known as the Pribnow box (Pribnow, 1975). Within the -35 region (TTGACA), the trimer TTG is strongly conserved, appearing at a frequency of 82, 84, and 79% for each base respectively (Hawley and McClure, 1983). At the -10 region (TATAAT), the first TA pair and the last T are also highly preserved. The last "T" in this hexamer is often referred to as the "invariant T". Although it is not absolutely required for promoter function, it is nevertheless present in all but four of the 112 promoters studied (Hawley and McClure, 1983).

A "prototypic" promoter should contain the -35 region, TTGACA, and the -10 region, TATAAT, with a distance of 17 bp in between. Generally, there is one promoter per gene. Sharing of promoters by more than one gene also occurs. On the other hand, there are genes which have more than one promoter such as the 'rrn' operons in E. coli (Glass, 1983; Newman and Morgan, 1977). They have at least two tandemly arranged promoters (Miura, Kruege, Itch, deBoer and Nomura, 1981; deBoer, Gilbert and Nomura, 1979). Recently two additional active promoters further upstream were identified in the 'rrnB' operon (Borus, Csordas-Toth, Kiss, Kiss, Torsk, Udverdy, Udverdy and Venetianer, 1983).

In prokaryotes, the termination site sequences consist of three common features (Rosenberg and Court, 1979; Rosenberg and Schmeissner, 1982): (a) an inverted repeat sequence precedes the termination site (b) G-C rich sequences variable in length (3-11 G-C pairs) are found in the inverted repeats and (c) a stretch of U residues are found in the terminus of the RNA transcript. These structural features are closely related to its function. After analysing more than 30 terminators of phage and bacterial origin, a simplified model for termination has been proposed (Platt, 1981).

The inverted repeat sequences found in the termination sites all have the potential ability to form stable base-paired stem and loop structures. This structure is essential for termination. The function of the stem and

144

loop structure may be to retard polymerase movement through the termination region (Greenblatt and Li, 1981). The stem length and the loop size are inconsistent due to the variation in both the length of the inverted repeats and the distance between the repeats. However, the G-C content of the inverted repeat is invariably rich. This is important because the high G-C content would stablize the RNA-DNA hybrid formed during transcription and might aid termination by impeding RNA polymerase movement on the template. Alternatively, the G-C base pair would also stablize the intramolecular stem and loop structure in RNA transcript that arise as a consequence of the dyad symmetry (Platt, 1981).

Most of the terminated RNA transcripts end with a run of four to eight consecutive U residues. These U's are thought to aid in the release of the transcript (Platt, 1981). It has been proposed that RNA polymerase (in the absence of other factors) requires both a stem and loop structure and a 3'-terminal stretch of U residues in the transcript in order to complete termination.

CHLOROPLAST PROMOTERS

Chloroplasts are probably of endosymbiotic origin and therefore possess the prokaryotic type machinery for protein synthesis (Kung, 1977). Generally, the structure of chloroplast genes resembles very closely those of prokaryotes with respect to control signals in the transcriptional cycle - promoters and terminators (Whitfield and Bottomley, 1983). In fact, chloroplast genes have been transcribed and translated faithfully *in vitro* using cell-free extracts of *E. coli* (Hartley, Wheerler and Ellis, 1975) or *in vivo* using various types of controlling signals including their own (Gatenby, Castleton and Saul, 1981; Erion, Tarnowski, Weissbach and Brot, 1981; Zhu, Lovett, Williams and Kung, 1984).

Recently, many chloroplast genes and their 5'-flanking regions have been sequenced and studied (Kung and Lin, 1985). In most cases, there are conserved promoter sequences in front of the chloroplast genes. Table I lists two groups of chloroplast promoters. The inclusion of promoters in the first group (A) was based on both structural and functional considerations. Chloroplast promoters must resemble those of prokaryotic promoters in structure and satisfy one of the following three functional criteria:

(a) these structures are protected by *E. coli* polymerase against DNase digestion; (b) these structures are determined by Sl mapping to be located in the 5' flanking region of a gene; and (c) these structures are active in initiating gene expression in either heterologous or homologous systems. All the promoters included in the second group (b) are defined structurally only. For example, a promoter can be identified by locating DNA sequences matching the structure of prokaryotic consensus sequence promoters proximal to a chloroplast gene. Most chloroplast promoters identified thus far fall into this group (b).

As in prokaryotes, chloroplast promoters also contain two conserved hexamers separated by a short stretch of about 17 bp. The first hexamer is TTGACA resembling the -35 region of prokaryotic promoter sequences, in which the trimer TTG is highly conserved. In the second hexamer, the conserved sequence of TATAAT is identical to the -10 region (McIntosh *et al,,* 1980). Within this hexamer, three bases (TA---T) are highly conserved. The first TA pair was present in all but few cases, in which the pair was either AA or TT (Table I). Similar to the prokaryotic promoters, the last "T" in this hexamer is also highly conserved, appearing in all but few of the 60 promoters compiled (Table I). The -35 and -10 region are separated by 11-24 bases which is very close to the allowed prokaryotic spacing of 15-21 bp (Pribnow, 1975). Overall, the distribution of bases in each position in these promoters of chloroplast genes is remarkably similar to that of prokaryotic promoters (Table II). However, no "prototypic" promoter has been identified for chloroplast genes (Table I).

In *E. coli,* secondary or tertiary promoters in the region upstream from the primary promoter sites for stable chloroplast RNA genes were also found (Table I). For example, there are two promoter sites reported for many tRNA genes (Deno, Kato, Shinozaki and Sugiura, 1982; Steinmetz, Krebbers, Schwartz, Gubbins and Bogorod, 1983). In most cases, the secondary promoters have -35 and -10 regions separated by 11-22 bp, and in a few cases only the individual isolated -10 regions were identified (Deno and Sugiura, 1984). The existence of multiple promoter sites was reported for maize and duckweed rRNA genes (Schwarz, Kossel, Schwarz and Bogorod, 1981; Keus, Dekker, vanRoon and Groot, 1983). The rRNAs (Tohdoh, Shinozaki and Sugiura, 1981), some tRNAs (Ohme, Komogashira, Shinozaki and Sugiura, 1985) and the β and ϵ (Zurawaski, Bottomley and Whitfield, 1982) genes are co-transcribed.

Table I   Compilation of chloroplast promoters from higher plants*

ORGANISMS AND GENES	PROMOTER SEQUENCES		
	-35 TTGACA		-10 TATAAT
E. coli Consensus Sequence			
(A) Structurally and functionally defined:			
N. tabacum rbcL	AAGTAAAAAGAAAAATTGGG	TTGCGCTATATATATGAAAGAGTA	TACAATAATGATGTATTTGGCAAATC
N. otophora rbcL	AAGTAAAAAGAAAAATTGGG	TTGCGCTATATATATGAAAGAGTA	TACAATAATGATCTATTTGGCAAATC
N. tabacum psbA	ATAGATCTACATACACCTTGG	TTGACACGAGTATATAAGTCATGT	TATACTGTTGAATAAAAGCCTTCCA
N. debneyi psbA	ATAGATCTACATACACCTTGG	TTGACACGAGTATATAAGTCATGT	TATACTGTTGAATAAAAGCCTTCCA
N. tabacum atpB	TCAGGTTCGAATTCCATAGAA	TAGATAATATGATGGGATTGTC	TATAATGATAGACAAATGAAAGACTT
N. otophora atpB	TCAGGTTCGAATTCCATAGAA	TAGATAATATGATGGGATTGTC	TATAATGATAGACAAATGAAAGACTT
N. tabacum 5S rRNA	GGTGTCCCCTCCAGTCAAGAA	TTGGGGCCTCACAATCACTAGCCAA	TATGCTTTTCTCTCATGCCTTTCTTC
16S rRNA	AGTTGTTCAAGAATAGTGGCG	TTGAGTTTCTCGACCCTTTGACT	TAGGATTAGTCAGTTCTATTTCTCGA
tRNA^gly	TGATTACCACAATTCCCCTGT	TCCGACAAAGTTGCATTGTA	TACAATAATCGGATTGTA
?	GCTGTGTTCGGGGGGAGTTA	TTGTCTATCGTTGGCCTCTATGG	TAGAATCAGTCGGGGGACCTGAGAGG
?	CGCACCATCGAAAACCGAATT	TTGCTGGTGGCTAACGTATACCCCTG	TAGCGTAACGTGACGACGTAACCAC
Maize rbcL	AAATAAAGATTAGGGTTTGGG	TTGCGCTATATCTATCAAAGAGTA	TACAATAATGATGATTGGTGAATC
atpB	AAATACTAAGAAAATTCTCTG	TTGACAGCAATCTATGCTTCACAG	TAGTATATATTTGTATATCGAAGTC

147

	16S rRNA	ATGGATAGGAGGCTTGTGGGA	TTGACGTGATAGGGTAGGGTTGGC	TATACTGCTGGTGGCGAACTCCAGGC
	tRNA Val(1)	TCCTATTTTCGATAGGACCCG	TTGACAATTGAATCCAATTTTCCCAT	TATTTGACTGTCCATAATAGTGCGGA
	tRNA Val(2)	AAGCCCGGAGGAGAGTGGGC	TTGCGTTTCTCGCCCCTTTGCCT	TAGGATTCGTTAATTCTCTTTCTCGA
	tRNA His	TCAGAATAAATAGAATAATAA	TGAATGGAAAAGAGAAAAATCCT	TTAGCTGGATAAGG
Spinach	rbcL	AAACCAACGGTTACCGGTTGGG	TTGCGCCATATATATGAAAGAGTA	TACAATAATGATGTATTTGGCGAATC
	atpB	TAAATAATTCGAAATTTACTC	TTGACAGTGGTATATGTTGTATATG	TATATCCTAGATGTGAAAATATGC
Wheat	rbcL	AGGATTAGGAATTAATTTGGG	TTGCGCTATATCTATCAAAGAGTA	TACAATAATTATGATTGGTAAATC
	atpB	AAATACTAATAAAATTCTTTG	TTGACAGCAATCTATGCTTCACAG	TAGTATATTTTGTATATATCGAAGTT
Mustard	psbA	ATCTTATCCATTTTACATTGG	TTGACATGGCTATATAAGTCATGT	TATACTGTTCAATAACAAGCTCTCAA
Pea	rbcL	CTCAAAAAAAAACGGTTGGG	TTGCGCCATACATATGAAAGAGTA	TAGAATAATGATGTATTTCCCAAA
	atpB	AAAAGATATTCTTGACC	TTGACAGTGATCTATGTTGTATATG	TAAATCCTAGATGTGAAAATCGGCAG

(B) Structurally defined

N. tabacum	tRNA Asn	AAGGGTATTAAATGAATGGAA	TTGGGATATATAGGATGGAA	TATAATGAAATAGAGCCACTTTGAGG
	tRNA His	AAAGAAGAGCTATATTCGAAC	TTGAATCTTTGTTTTCTAATTTA	AATAATGTAAAAACGAATGTAAGTA
	tRNA Met	TGTATAAATGGGCTATTCTAT	TTGTACAGATAGGGTGGAGGGGCGCA	TTTAATCCTTGTTTATCTATTAGTTT
	tRNA Pro	CGGGTTCTCGTATTTATATATT	TTGTATATAATTGTATATAAGTATTTTCTA	TATAATCTATAAGAGAAGTCTTTTCC
	tRNA Trp	ATCAATTGAGATCGCCTCAAA	TTGGACATAATCTTTGATTTT	TATCATGCTATTCTAGTATATGCATA
	tRNA Arg(1)			TACAATTCCAAAATTCTTTCACATC

tRNA^{Arg}(2)			TAAAATACGAAAAAAATCAGAATG
tRNA^{Val}(1)	GCTCAAAGAGATCAAAGATTG	TTGATGTTGGATCATGGAATATT	TATCTTGACAAGAATTTATCTACATG
tRNA^{Val}(2)	TTGGATCATGGAATATTTATC	TTGACAAGAATTTATCTACATGA	TAAAATATGTATCACAAGCACTA
CS19	CCCCTGGGGTTATCCTGCAC	TTGGAAGAAGAAGTAGAAAAAGGAATAAA	TATAGTGATAATTTGATTCTTCGTCG
tRNA^{Gly}(1)	TGATTACCACAATTCCCCTGT	TCGACAAAAGTTGCATTGTA	TACAATAATCGGATTGTA
tRNA^{Gly}(2)	AGAGAATATGTGTCCCGGCAC	TGCACAAAAAGATCCGGTTATA	TATCATATATGTGGGTACATATTGTG
CF. III	CCCTTCTAGATGTTCGACGC	TTGATTCTCGAATAGGATTGAATC	TAAGATGAATGCTTGGTTTACGTTAT
Maize tRNA^{Leu}(1)	TAATGAATTCAATGA	TTCAAAAAAACTAAGAGATGGA	TTAAATTATACAAGGAATCCTGGTTT
tRNA^{Ser}(1)	GAGTTAGTAGATCATTCATA	TAGCTATGTTCTATTTGTAGGAA	TAAAATAGGGGATTGGGCTGT
tRNA^{Ser}(2)	CAGGAATACGAAAACTCGCTA	TTCACTCAGTTTATTTTCCATAA	TAAGATTATGTA
tRNA^{Val}	TGGCATTAGAGAATATTCATC	TTGACAAGAAATTATCTATATGT	TAAGATATCTCTGAC
tRNA^{Leu}(1)	AAGACTCCACCT	TGTCATATATTCCATATATCACA	TTCGATAGATATCATATTCATGGAAT
tRNA^{Leu}(2)	AGACTCCACCTTTGTCATATA	TTCCATATATCACATTCGATAGA	TATCATATTCATGGAATACGATTCAC
tRNA^{Met}	CATACCAATAACGGAGCGGTA	TTGCTTATAAAAGGATTCAATC	TATAATCGATCGAAGTAATGGGGCTT
tRNA^{Phe}	TTGATTTTTTAGTCCCTTTAA	TTGACATAGATGCAAATACTTTAC	TAAGATGATGCACAAGAAAGG
tRNA^{Thr}(1)	CTATCTAAGTGGAACTTCCAA	TTTAGAACTAGTTAATAAC	TAAGATTAATAATTAAGATCTGACAT
tRNA^{Thr}(2)	GAACTTCCAATTAGAACTAG	TTAATAACTAAGATTAATAAT	TAAGATCTGACATTTTACAGATTCCC
D2(1)	TAATATAGAAAACGATTTTTT	TTGATTCACGAACAAGATTCAAGAA	TAATCTTATTTGATAAAGCAGAGTA
D2(2)	GTTAATGGATTGACCTAGAT	TAGATATCAATCGACAAAAAAA	TAATTTTTCTATTCGAAACCCAGTCG

Spinach	P680	AGACGATGCTATCAACTCCGA	TTGCGTATTGCTACTTATCGAGTA	TAGAATAGATTGTTTCTCTTTGTTC
	psbA	ATAGATCTCACTAGATATTGG	TTGACACGGGCATATAAGGCATGT	TATACTGTTGAATAACAATCTTTAA
S. nigrum	psbA	ATAGATCCAGATACAGCTTGG	TTGACACGAGTATATAAGTCATGT	TATACTGTTGAATAACAAGCCTTCCA
Soybean	psbA	TACTATGGATATTGGTATTGG	TTGACACTGGTATATAAGTCATGT	TATACTGTTGAATAACAAGTCCTCAA
Duckweed	16S rRNA	ATGAATAAGAGGCTCGTGGGA	TTGACGTGATAGGGTAGGGATGGC	TATATTGCTGGGAGCCGAACCTCCAG
	5s rRNA	GGTGTCCCTCCAGTCAAGAA	TTGGGGCCTCACAATCACTAGCCAA	TATGAATATGCTTTTCTCTCATGACT
Broadbean	rbcL	GACTCAAAAAAACGGTTGGG	TTGCGCCATACATATGAAACAGTA	TAGAATAATGATGTATTTGCCAAATC
	atpB	AAAGTTCAGGTTCGAATTACA	TAGATAATATAGATAGTATTGTC	TATAATCTAGAATGATAAACAAATGA
	tRNA Glu	GAATCATATCATTCCATTATA	TTGACAATTTCAAAAAACTGTTCA	TACTATGAACATAGTAGAATGGAAAT
	rRNA Thr(1)	TGTACTAAACTCATCTTCATA	TTGGCTGATTCCGTATTGGGGAA	TTTACTCAAACGCC
	tRNA Thr(2)	ATATATATCTATTTCGTCAGA	TTGATATACCAATTTTGTATATATC	TATTTTGTATATCTATCTATAATAAT

*: from Kung and Lin (1985)

Chloroplast promoters are structurally **well**-defined. They are present in front of almost every sequenced chloroplast gene. Therefore, they are likely to be the active promoters regulating the expression of chloroplast genes. The functional analyses of a few selected promoters support this claim.

Table II   Distribution of bases at each position in promoters of chloroplast genes

Position	Distribution											
	T	T	G	A	C	A	T	A	T	A	A	T
A	0	5	2	35	7	30	1	55	14	36	40	1
T	58	49	1	4	13	12	59	5	27	7	7	56
G	0	2	51	6	14	7			9	12	2	1
C	0	2	4	13	24	9			10	5	11	2
Chloroplast (%)*	100	84	88	60	41	52	98	92	45	60	67	93
Prokaryotes (%)**	82	84	79	64	54	45	81	95	44	59	51	96

* Calculated from Table I
** Calculated from Hawley and McClure (1983)

Functional Analysis of some structurally defined chloroplast promoters

Recently, the gene fusion technique (Rosenberg *et al*., 1983) has been applied as a functional assay to study the putative chloroplast promoters (Kong *et al.*, 1984). Gene fusion makes use of a promoter-probe plasmid in the isolation, identification and characterization of promoters. The plasmid (*eg.*, pKO1) contains a promoterless *E. coli galK* gene. Insertion of a promoter in the proper orientation results in *galK* expression. Moreover, the precise levels of expression can be monitored by the *galK* enzyme assay (Rosenberg *et al.*, 1983).

This functional assay of chloroplast promoters in the pKO1 system established a certain structure-function relationship of the sequences tested. However, this relationship is not proof that these promoters are authentic chloroplast promoters. Such proof must come from experiments in which it can be demonstrated that deletion of certain sequences in the promoter region causes the failure of homologous chloroplast RNA polymerase to initiate transcription. The results of such experiments have recently been reported (Link, 1984). Using plasmids with sequential deletion in the 5'-flanking region of the *psb-A* gene, and a homologous chloroplast to initiate transcription. The results of such experiments have recently been reported (Link, 1984). Using plasmids with sequential deletions in the 5'-flanking region of the *ps-A* gene, and a homologous chloroplast extract. Link (1985) demonstrated that the upstream region containing the sequences of TTGACA and TATACT, which resemble the prokaryotic -35 and -10 regions, is required for efficient *in vitro* transcription.

Using a similar technique Chua and his co-workers (Orozco, Hanley-Bowdoin, Poulsen and Chua, 1984) have demonstrated that increasing the distance between the -35 and -10 regions of maize rbcL promoter from 18 bp to 20 bp with an AT base pair insertion reduced the level of transcription drastically. It is clear that the -35, the -10 regions and the distance between them are involved in the modulation of the efficiency of chloroplast promoters. The evidence obtained from the functional assays in the homologous system also demonstrated that many of the structurally well-defined promoters are indeed functionally active chloroplast promoters (Orozco *et al.*, 1985; Hanley-Bardoin, Orozco and Chua, 1985).

Interchangeability of prokaryotic and chloroplast promoters

In view of the extensive similarities between prokaryotic and chloroplast systems it is not surprising that promoters from chloroplasts and prokaryotes are interchangeable. After the successful expression of the spinach *rbcL* gene *in vitro* in coupled transcription and translation systems using cell-free extracts of *E. coli* (Bottomley and Whitfield, 1979), this gene has been expressed in *E. coli* and *B. subtilis* under control of its own (Gatenby *et al.*, 1981; Erion *et al.*, 1981; Zhu *et al.*, 1984) or bacterial promoters (Zhu *et al.*, 1984; Steinmetz *et al.*, 1983).

Amplification of the *rbcL* gene by using fused lambda and *cat* promoters in prokaryotic systems was also reported (Rosenberg *et al.*, 1983; Zhu *et al.*, 1984). The expression of maize *rbcL* gene was amplified under a temperature sensitive lambda promoter. Likewise, when the promoterless *rbcL* gene from *Chlamydomonas* was placed under the control of the *cat* promoter its

expression in *B. subtilis* was induced with the chloramphenicol (Zhu *et al.*,
1984). Conversely, the *galK* and *cat* genes have been expressed and
amplified under the control of chloroplast promoters (Kong *et al.*, 1984;
Gatenby and Castleton, 1982). Thus, the structural similarity of chloroplast and prokaryotic promoters is further supported by their functional
interchangeability.

Prediction of promoter strength from DNA sequences
If some specific DNA sequences of a promoter are correlated with its
strength, such correlation may exist in chloroplast promoters. Available
evidence suggests that both *psbA* and *rbcL* promoters are strong signals.
Both contain highly conserved bases in the sequence; the TTG in the -35
region, and the first TA pair and the last T in the -10 region, with a
near optimal distance of 18 bp (Table III). In fact, the *psbA* promoter is
only a single base away from being "perfect". Both promoters are totally
conserved in the region from -10 to -35 among plant species; being
identical in all seven plant species sequenced to date for *psbA* and *rbcL*.
There is no such feature found in any other promoters, such as *atp β*gene
(Table III).

    Both *psbA* and *rbcL* promoters are AT-rich (70%) and possess a TATA-like
element between the -35 and -10 regions. This eukaryotic element is also
functional in chloroplasts (Link, 1984). In addition to the unique structural features, there are some obvious functional properties that support
the notion that they are strong promoters. For example, *psbA* and *rbcL*
mRNAs are the most abundant mRNA species in chloroplasts (Mattoo, Foffman-
Falk, Marder and Edelman, 1984). Both proteins are synthesized very
rapidly in chloroplasts. It is conceivable that the structural features
are responsible for this functional consequence.

    Many elements in the promoter sequences will certainly affect the
promoter strength. Already there is evidence to indicate that besides the
three well-defined elements (-35, -10 regions and a spacer), the following
features are important: (a) the nucleotide sequence of the spacer; (b)
the presence of secondary promoter sites; (c) the overall AT richness of
the promoter region as well as the regions upstream and (d) any particular
combinations of these features. By considering all these features together,
the strength of a promoter may be predicted.

Table III The *psbA* and *rbcL* promoters from different plant species (from Table I)

Plant Species	psbA			rbcL			atpB	
	-35	TATA	-10	-35	TATA	-10	-35	-10
S. nigrum	TTGACACGAGTATATAAGTCATGTTATACT							
Mustard	TTGACATGGCTATATAAGTCATGTTATACT							
Soybean	TTGACACTGGTATATAAGTCATGTTATACT							
N. debneyi	TTGACACGAGTATATAAGTCATGTTATACT							
N. tabacum	TTGACACCGATATATAAGTCATGTTATACT			TTGCGCTATATATGAAAGAGTATACAAT			TAGATAATATGGATGGGATTGTC	TATAAT
Spinach	TTGACACGGGCATATAAGGCATGTTATACT			TTGCGCCATATATGAAAGAGTATACAAT			TTGACAGTGGTATATGTTGTATATG	TATATC
N. otophora				TTGCGCTATATATGAAAGAGTATACAAT			TAGATAAATGGATGGGATTGTC	TATAAT
Broadbean				TTGCGCCATACATGAAACAGTATACAAT			TAGATAATAAGATAGTATTGTC	TATAAT
Pea				TTGCGCCATACATGAAAGAGTATACAAT			TTGACAGTGATCTATGTtGtATATG	TAAATC
Maize				TTGCGCTATATCACTCAAAGAGTATACAAT			TTGACAGCAATCTATGCTTCACAG	TAGTAT
Wheat				TTGCGCTATATCTATCAAAGAGTATACAAT			TTGACAGCAATCTATGCTTCACAG	TAGTAT

154

CHLOROPLAST TERMINATORS

Many chloroplast genes and their 5' and 3' flanking regions have been
sequenced (Morris and Herrmann, 1984; Deno, Shinozaki and Sugiura, 1985).
In the 5' flanking region there are conserved promoter sequences in front
of almost every chloroplast gene analysed. Similarly, in the 3' flanking
region the most obvious common feature of many sequenced chloroplast genes
is the presence of complementary sequences capable of forming a stem and
loop structure similar to that found in prokaryotes (Platt, 1981).
Such a feature can be seen in *rbcL* from spinach (Zurawski *et al.*, 1981)
maize (McIntosh *et al.*, 1980) *Nicotiana* (Shinozaki and Sugiura 1981)
Petunia (Bovenberg *et al.*, 1984) as well as in *psbA* from spinach (Zurawski
et al., 1982) *Nicotiana* (Zurawski *et al.*, 1982; Sugita and Sugiura, 1984),

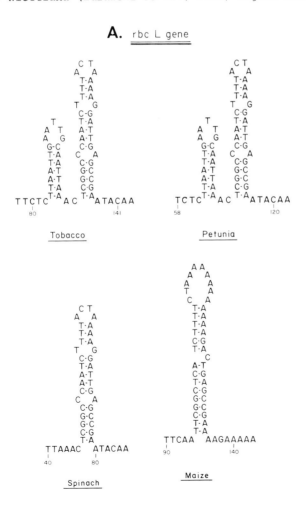

A. rbc L gene

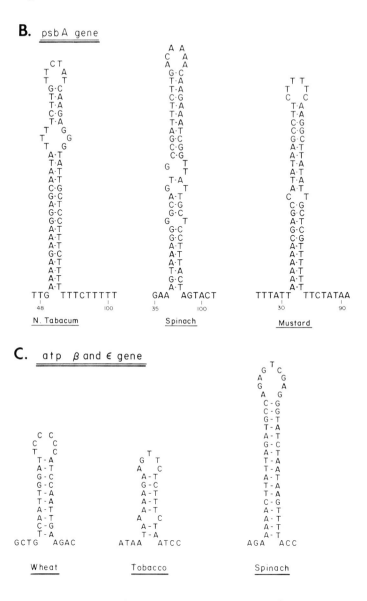

Fig. 1 - Proposed stem and loop structures at the 3' end of
chloroplast genes from higher plants. They represent
the putative terminators for the respective genes. A
rbcL gene Zurawski *et al.*, 1981; McIntosh *et al.*, 1980;
Shinozaki *et al.*, 1981; Bovenberg *et al.*, 1984);
B. *psbA* gene (Zurawski *et al.*, 1982; Sugita *et al.*, 1984;
Link and Langridge, 1984); and C. *atp* and gene (Shinozaki
et al., 1983; Whitfield *et al.*, 1983; Howe *et al.*, 1985).

mustard (Link and Langridge 1984), and others (Shinozaki, Deno and Sugiura, 1983; Whitfield, Zurawski and Bottomley, 1983; Howe, Fearnley, Walker, Dyer and Gray, 1985) (Fig. 1). Because of the structural resemblance, it is likely that such structures will function as recognition signals for termination of transcription by chloroplast RNA polymerase. The information begins to appear that they indeed function as terminators at least in the heterologous system.

A. *In vivo* analysis of structure and function of chloroplast terminators

When *N. otophora* ct-DNA was digested with Sau3A restriction enzyme and subsequently inserted into the terminator-probe plasmid pDR720 at the *Bam*HI site between the *trp* promoter and the *galK* gene, reduction of the *galK* gene expression on McConky plates was observed (Akeda, Mong, Kolsuk, Lovett, Hirai and Kung, 1985). These inserts are of ct-DNA origin. When a small insert from ct-DNA was sequenced, it contains 429 bp with a short inverted repeat capable of forming a stem-loop structure. This terminator consisted of one 10 bp inserted repeat sequences followed by a run of 3-6 residues resembling very closely that of prokoryotic terminators (Platt, 1981). Thus, a structure-function relationship of chloroplast terminator as tested in the prokaryotic system is established. This 429 bp ct-DNA fragment was transferred from the pKO1 system to pSM 301 containing the *cat* gene. Chloramphenicol acetyltransferase (CAT) assays (Akada *et al.*, 1985 clearly demonstrated that this ct-DNA fragment also functions as an efficient terminator by reducing the level of CAT activity. This reduction is caused by termination of *cat* gene expression as demonstrated by *S1* mapping (Akada *et al.*, 1985).

The structure-function relationship of chloroplast terminators was substantiated by the functional analysis of the structurally defined *Nicotiana rbcL* terminator. A 380 bp *XbaI* fragment at the 3' end of the *rbcL* gene from *N. tabacum* and *N. otophora* containing the stem-loop structure exhibited termination function for both *GalK* and *Cat* genes (Akada *et al.*, 1985).

Both terminators are very efficient as tested in the prokaryotic system. Analysis of the *cat* gene activity showed that they reduce the CAT levels by over 98%. Both terminators function effectively regardless of the presence or absence of Rho factors.

DISCUSSION

The aim of this article is to review and to present the case that promoters and terminators of chloroplast and prokaryotic origin are structurally similar. Some chloroplast promoters are not only structurally indistinguishable with those of prokaryotes. The control of transcription in chloroplast genes is prokaryotic in nature. These observations support the view that chloroplasts are of endosymbiotic origin.

The evidence now available clearly establishes a structure-function relationship of chloroplast promoters and terminators. When the promoter and terminator-active frangments of ct-DNA are sequenced, they exhibit the prokaryotic promoter and terminator features (Kong *et al.,* 1984; Akada *et al.,* 1985). Similarily, when the promoter – and terminator-like sequences are functionally analyzed, they act as promoter and terminators. The pKO system provides a rapid method for screening chloroplast promoter and terminator sequences. However, it should be pointed out that the results obtained from the pko system should be verified by using chloroplast enzymes, since the prokaryotic system is more permissive than the chloroplast system, and can recognize and initiate transcription at sequences within the chloroplast genome which are not authentic chloroplast promoters (Hanley-Bowdoin *et al.,* 1985).

REFERENCES

Akada, S., Mongkolsuk, S., Lovett, P.S., Hirai, A. and Kung, S.D. (1985). *Nicotiana* chloroplast genome XI: *In vivo* analysis of the structure and function of terminators. Gene: submitted.

de Boer, H.A., Gilbert, S.F,, and Nomura, M. (1979). DNA sequences of promoter regions for rRNA operons rrnE and rrnA in *E. coli*. *Cell 17:* 201 – 09.

Boros, I., Csordas--Toth, E., Kiss, A., Kiss, I., Torok, I., Udvardy, A., Udvardy, K., and Venetianer, P. (1983). Identification of two new promoters probably involved in the transcription of a ribosomal RNA gene of *Escherichia coli*. Biochim. Biophys. *Acta. 739:* 173 – 80.

Bottomley, W. and Whitfeld, P.R., (1979). Cell-free transcription and translation of total spinach chloroplast DNA. *Eur. J. Biochem. 93:* 31 – 39.

Bovenberg, W.A., Koes, R.E., Kool, A.J. and Nijkamp, J.J. (1984). Physical mapping, nucleotide sequencing and expression in *E. coli* minicells of the gene for the large subunit of ribulose bisphosphate carboxylase from *Petunia hybrida*. *Curr. Genet. 8:* 231 - 41.

Deno, H., Kato, K., Shinozaki, K. and Sugiura, M. (1982). Nucleotide sequences of tobacco chloroplast genes for elongator tRNA^{Met} and tRNA^{Val} (UAC) gene contains a long intron. *Nucleic Acids Res. 10:* 7511 - 20.

Deno, H., Shinozaki, K. and Sugiura, M. (1985). Structure and transcription of a tobacco chloroplast gene coding for subunit III of proton-translocating ATPase. *Gene 32:* 195 - 201.

Deno, H. and Sugiura, M. (1984). Chloroplast tRNAG^{ly} gene contains a long intron in the D stem: nucleotide sequences of tobacco chloroplast genes for tRNA^{Gly} (UCC) and tRNA^{Arg} (UCU). *Proc. Natl. Acad. Sci. USA 81:* 405 - 8.

Erion, J.L., Tarnowski, J., Weissbach, H. and Brot, N. (1981). Cloning, mapping, and *in vitro* transcription-translation of the gene for the large subunit of ribulose-1, 5-bisphosphate carboxylase from spinach chloroplasts. *Proc. Natl. Acad. Sci. USA 78:* 3459 - 63.

Gatenby, A.A., Castleton, J.A. and Saul, M.W. (1981). Expression in *E. coli* of maize and wheat chloroplast genes for the large subunit of ribulose bisphosphate carboxylase. *Nature 291:* 117 - 21.

Gatenby, A.A. and Castleton, J.A. (1982). Amplification of maize ribulose bisphosphate carboxylase large subunit synthesis in *E. coli* by transcriptional fusion with the lambda N operon. *Mol. Gen. Genet. 185:* 424 - 29.

Glass, R.E. (1983). *Gene function*. Croom Helm Ltd. London.

Greenblatt, J. and Li, J. (1981). The nusA gene protein of *E. coli:* Its identification and a demonstration that it interacts with the gene N transcription anti-termination protein of bacteriophage lambda. *J. Mol. Biol. 147:* 11 - 23.

Hanley-Bowdoin, L., Orozco, E.M. Jr. and Chua, N.H. (1985). Transcription of chloroplast genes by homologous and heterologous RNA polymerases. In: *Photosynthetic apparatus*. Cold Spring Harbor Press. (in press).

Hartley, M.R., Wheerler, A. and Ellis, R.J. (1975). Protein synthesis in chloroplasts V: Translation of messenger RNA for the large subunit of fraction I protein in a heterologous cell-free system. *J. Mol. Biol. 91:* 67 - 77.

Hawley, D.K. and McClure, W.R. (1983). Compilation and analysis of
   Escherichia coli promoter DNA sequences. *Nucleic Acids Res.* 11: 237 - 55.

Howe, C.J., Fearnley, I.M., Walker, J.E., Dyer, and Gray, J.C. (1985).
   Nucleotide sequences of the genes for the alpha, beta and epsilon subunits of wheat chloroplast ATP synthase. *Plant Mol. Biol.* 4: 333 - 45.

Keus, R.J.A., Dekker, A.F., van Roon, M.A. and Groot, G.S.P. (1983). The
   nucleotide sequences of the regions flanking the genes coding for 23S,
   16S and 4.5S ribosomal RNA on chloroplast DNA from *Spirodela oligorbiza*.
   Nucleic Acids Res. 11: 6465 - 74.

Kong, X.F., Lovett, P.S. and Kung, S.D. (1984). The *Nicotiana* chloroplast
   genome IX: identification of regions active as prokaryotic promoters in
   Escherichia coli. *Gene 31*: 23 - 30.

Kung, S.D. (1977). Expression of chloroplast genomes in higher plants.
   Ann. Rev. Plant Physiol. 28: 401 - 37.

Kung, S.D. and Lin, C.M. (1985). Surveys and summaries on chloroplast
   promoters from higher plants. *Nucleic Acids Res.*: (in press).

Link, G. (1984). DNA sequence requirements for the accurate transcription
   of a protein-coding plastid gene in a plastid *in vitro* system from
   mustard (*Sinapis alba L.*). *EMBO. J. 3*:1697 - 1704.

Link, G. and Langridge, U. (1984). Structure of the chloroplast gene for
   the precursor of the Mr 32,000 photosystem II protein from mustard
   (*Sinapis alba L.*). *Nucleic Acids Res. 12*: 945 - 58.

Mattoo, A., Foffman-Falk, H., Marder, J.B. and Edelman, M. (198 4).
   Regulation of protein metabolism: Coupling of photosysthetic electron
   transport to *in vivo* degradation of the rapidly metabolized 32-kilodalton
   protein of the chloroplast membranes. *Proc. Natl. Acad. Sci. USA 81*:
   1380 - 84.

McIntosh, L., Poulsen, C. and Bogorad, L. (1980). Chloroplast gene sequence
   for the large subunit of ribulose bisphosphate carboxylase of maize.
   Nature 288: 556 - 60.

Miura, A., Kruege, J.H., Itoh, S., de Boer, H.A. and Nomura, M. (1981).
   Growth-rate-dependent regulation of ribosome synthesis in *E. coli*:
   expression of the lacZ and galK genes fused to ribosomal promoters.
   Cell 25: 773 - 82.

Morris, J. and Herrmann, R.G. (1984). Nucleotide sequence of the gene for
   the P680 chlorophyll a apoprotein of the photosystem II reaction centre

from spinach. *Nucleic Acids Res.* 12: 2837 - 50.

Nomura, M. and Morgan, E. (1977). Genetics of bacterial ribosomes. *Ann. Rev. Genet.* 11: 297 - 347.

Ohme, M., Kamogashira, T., Shinozaki, K. and Sugiura, M. (1985). Structure and cotranscription of tobacco chloroplast genes for tRNA^{Glu} (UUC), tRNA^{Tyr} (GUA) and tRNA^{Asp} (GUU). *Nucleic Acid Res.* 13: 1045 - 56.

Orozco, E.M. Jr., Hanley-Bowdoin, L., Poulsen, C. and Chua, N.H. (1984) Identification of spinach and maize plastid DNA sequences required for transcription initiation *in vitro* the *rbcL* and *atpB* genes. In: *Plant Genes: structure, expression, mobility.* Abstract. Twenty-second Harden Conference. The Biochemistry Society. 1984. L27.

Orozco, E.M. Jr., Mullet, J.E. and Chua, N.H. (1985). An *in vitro* system for accurate transcription initiation of chloroplast protein genes. *Nucleic Acids Res.* 13: 1283 - 1302.

Platt, T. (o981). Termination of transcription and its regulation in the tryptophan operon of *E. coli*. *Cell* 24: 10 - 23.

Pribnow, D. (1975). Nucleotide sequence of an RNA polymerase binding site at an early T7 promoter. *Proc. Natl. Acad. Sci. USA* 72: 784 - 89.

Roberts, J. (1969). Termination factor for RNA synthesis. *Nature 224*: 1168 - 74.

Rosenberg, M., Chepelinsky, A.B. and McKenny, K. (1983). Studying promoters and terminators by gene fusion. *Science 222*: 734 - 39.

Rosenberg, M. and Court, D. (1979). Regulatory sequences involved in the promotion and termination of RNA transcription. *Ann. Rev. Genet.* 13: 319 - 53.

Rosenberg, M. and Schmeissner, U. (1982). In: *Interaction of Translational and Transcriptional Controls in the Regulation of Gene Expression.* (eds. M. Grunsberg-Manazo and B. Safer) Elsevier Sci. Publication Co. ppl.

Schwarz, Zs., Kossel, H., Schwarz, E., and Bogorad, L. (1981). A gene coding for tRNA^{Val} is located near 5'-terminus of 16S rRNA gene in *Zea mays* chloroplast genome. *Proc. Natl. Acad. Sci. USA 78*: 4748 - 52.

Shinozaki, K., Deno, A.K., and Sugiura, M. (1983). Overlap and co-transcription of the genes for the beta and epsilon subunits of tobacco chloroplast ATPase. *Gene 24*: 147 - 55.

Shinozaki, K. and Sugiura, M. (1981). The nucleotide sequence of the tobacco chloroplast gene for the large subunit of ribulose-1,

5-bisphosphate carboxylase/oxygenase. *Gene 20:* 91 - 102.

Steinmetz, A.A., Krebbers, E.T., Schwarz, Zs., Gubbins, E.J. and Bogorad, L. (1983). Nucleotide sequences of five maize chloroplast transfer RNA genes and their flanking regions. *J. Biol. Chem. 258:* 5503 - 11.

Sugita, M. and Sugiura, M. (1984). Nucleotide sequence and transcription of the gene for the 32,000 dalton thylakoid membrane protein from *Nicotiana tabacum. Mol. Gen. Genet. 195:* 308 - 313.

Tohdoh, N., Shinozaki, K. and Sugiura, M. (1981). Sequence of a putative promoter region for the rRNA genes of tobacco chloroplast DNA. *Nucleic Acids. Res. 9:* 5399 - 5406.

Travers, A.A. (1984). Conserved features of coordinately regulated *E. coli* promoters. *Nucleic Acids Res. 12:* 2605 - 18.

Whitfeld, P.R. and Bottomley, W. (1983). Organization and structure of chloroplast genes. *Ann. Rev. Plant Physiol. 34:* 279 - 310.

Whitfeld, P.R., Zurawski, G. and Bottomley, W. (1983). Features revealed by the sequencing of chloroplast genes highlight the prokaryote-like nature of these organelles. In: *Manipulation and Expression of Genes in Eukaryotes.* (eds. Nagley, P., Linnane, A.W., Peacock, W.J. and Pateman, J.A.) Academic Press. New York. 1983. P.247 - 54.

Yanofsky, C. (1981). Attenuation in the control of expression of bacterial operons. *Nature 289:* 751 - 58.

Zhu, Y.S., Lovett, P.S., Williams, D.M. and Kung, S.D. (1984). *Nicotiana* chloroplast genome VII: expression in *E. coli* and *B. subtilis* of tobacco and *Chlamydomonas* chloroplast DNA sequences coding for the large subunit of ribulose diphosphate (RuBP) carboxylase. *Theor. Appl. Genet. 67:* 333 - 36.

Zurawski, G., Bohnert, H.J., Whitfeld, P.R. and Bottomley, W. (1982). Nucleotide sequence of the gene for the M_R 32,000 thylakoid membrane protein from *Spinacia oleracea* and *Nicotiana debneyi* predicts a totally conserved promary translation product of M_R 38,950. *Proc. Natl. Acad. Sci. USA 79:* 7699 - 7703.

Zurawski, G., Bottomley, W. and Whitfeld, P.R. (1982). Structure of the genes for the B and E subunits of spinach chloroplast ATPase indicates a dicistronic mRNA and an overlapping translation stop/start signal. *Proc. Natl. Acad. SXi. USA 79:* 6260 - 64.

Zurawski, G., Perrot, B., Bottomley, W. and Whitfeld, P.R. (1981). The structure of the gene for the large subunit of ribulose-1, 5-bisphosphate carboxylase from spinach chloroplast DNA. *Nucleic Acids Res. 9:* 3251 - 70.

7 Recombinogenic sequences in chloroplast DNA

C. J. Howe

ABSTRACT

A number of lines of evidence which indicate the occurrence of recombination and related processes, such as gene conversion, in ct-DNA will be briefly reviewed. The possible role of homologous recombination in bringing about an inversion in the ct-DNA of a number of monocots will be discussed, and some implications of these findings for plant genetic manipulation will be considered.

INTRODUCTION

Recombination is a widespread phenomenon of rearrangement of DNA sequences. It will be important to know the extent and mechanism of this process in chloroplasts, since this will affect our understanding of a wide range of problems such as chloroplast DNA evolution, the induction of chloroplast DNA mutations and maternal inheritance. In addition, it will influence attempts at artificial genetic manipulation of the chloroplast.

EVIDENCE OF RECOMBINATION IN CHLOROPLASTS

Such a wide range of organisms, both prokaryotic and eukaryotic, exhibit genetic recombination that it would be most surprising if it did not take place in chloroplasts as well (see Chapter 10). However, there are several experimental approaches which indicate the occurrence of recombination and also related processes, such as gene conversion. Classical genetic recombination was used by Sager and Ramanis (1970) to construct a map for non-Mendelian chloroplast genes in *Chlamydomonas*. This depended upon UV-irradiation of the female parent before mating to bring about the formation biparental progeny, i.e. progeny carrying chloroplast genomes from both parents rather than progeny showing strictly uniparental inheritance.

Using appropriate genetic markers, recombination between the two parental chloroplast genomes could be demonstrated. In addition, there is evidence from similar experiments for gene conversion in *Chlamydomonas* chloroplasts (Gillham, 1978).

In higher plants, the evidence for recombination comes from more direct analysis of the ct-DNA. Kolodner and Tewari (1979) analysed denatured and partially renatured ct-DNA by electron microscopy. They confirmed that the monomeric ct-DNA molecules from spinach, lettuce and maize contained a large inverted repeat sequence separating two single-copy regions. Pea ct-DNA did not show the inverted repeat structure. All four plants were found to contain dimeric chloroplast DNA molecules whose structures after self-renaturation were consistent with their being "head-to-tail" dimers. These could be formed by recombination between ct-DNA molecules at homologous regions anywhere along their length. Kolodner and Tewari (1979) also observed structures which they termed "head-to-head" circular dimers. They interpreted these as resulting from *inter*-molecular recombination across opposing halves of the inverted repeats; consistent with the fact that they were not observed in pea. Such forms could also be obtained by a combination of *intra*-molecular recombination across the two halves of the repeat and *inter*-molecular recombination across any homologous points in the chromosomes.

The occurrence of recombination across the elements of the inverted repeat has also been inferred on the basis of restriction enzyme analysis of ct-DNA. Recombination between the two halves will generate a molecule in which the relative orientations of the single copy regions has been "flipped". Digestion of ct-DNA with an enzyme which does not cleave within the repeats allows the two forms to be distinguished. The presence of roughly equal quantities of these two forms has now been demonstrated in a number of ct-DNAs, including those of *Cyanophora, Phaseolus vulgaris,* soybean and *Osmunda* (Bohnert and Loffelhardt, 1982; Palmer, 1983; Palmer, Osario, Watson, Edwards, Dodd and Thompson, 1984). That nearly equal amounts of both forms are found within a single plant indicates that the recombination is frequent within the lifetime of the plant, otherwise stochastic fluctuations might be expected to lead to an imbalance (unless some subtle mechanism operates to maintain equal proportions).

Intramolecular recombination is also likely to be the cause of such inversions and other chromosomal rearrangements as have taken place during the evolution of ct-DNA. In general, the ct-DNA of angiosperms is remarkably similar in gene organisation. There are, however, some notable exceptions. For example, during the evolution of certain monocotyledons, such as maize, wheat and barley, but not *Spirodela*, simple inversions have taken place (Palmer and Thompson, 1982; Howe, Bowman, Dyer and Gray, 1983; Oliver, 1984; de Heij, Lustig, Moeskops). A rather different rearrangement has taken place in rice (Hirai, Ishibashi, Morikami, Jwatsuki, Shinozaki and Sugiura, 1985). An inversion has also occurred in the evolution of *Oenothera* (Bovenberg, Bisanz and Groot, 1983) ct-DNA (Herrmann, Westhoff, Alt, Winter, Tittgen, Bisanz, Sears, Nelson, Hurt, Hanuska, Viebrock and Sebald, 1983), and a very similar one is found in mung bean. Many legumes, generally those that have lost one of the two inverted repeats, show a very complex set of rearrangements (Palmer and Thompson, 1982). In addition, comparison of the closely related plastid types of *Oenothera* has shown that certain sections of the chloroplast genome have been particularly subject to length changes. This is also true for related species within the genera *Nicotiana, Triticum* and *Aegilops* (Kung, Zhu and Shen, 1982; Saltz, Herrmann, Peleg, Lavi, Izhar, Frankel and Beckmann, 1984; Ogihara and Tsunewaki, 1982; Bowman, Bonnard and Dyer, 1983). These changes may be the result of unequal crossing over generating duplications and deletions. Particularly striking is the fact that the same regions of the chromosome have been involved in a number of different rearrangements. This is shown in Fig. 1. Certain regions of ct-DNA appear to be acting as "hotspots", and it is tempting to draw an analogy with the "chi" sequence of *E. coli*, which stimulates recombination (Stahl, Crasemann and Stahl, 1975). Clearly, molecular analysis of such hotspots may yield valuable clues to the mechanism of recombination.

Paradoxically, a further piece of evidence for recombination processes is furnished by the homogeneity of ct-DNA molecules (apart from the "flip-flop" described earlier) within a plant. Furthermore, species comparisons of ct-DNA within the Solanaceae indicate that fixation of a mutation in one half of the inverted repeat is accompanied by a parallel mutation in the other (Fluhr and Edelmann, 1981). These observations suggest an active process of gene conversion (which is usually associated with recombination) to maintain molecular homogeneity. An exception to this homogeneity is found in albino wheat plants regenerated from pollen by anther culture. Such plants contain a heterogeneous collection of partially deleted ct-DNA molecules (Day and Ellis, 1984). It is possible that this may reflect a programmed destruction of ct-DNA in the male plant, bringing about maternal inheritance. Deletions of ct-DNA in *Chlamydomonas* and *Euglena*, often involving rDNA can also be brought about by treatment with appropriate agents, such as 5-fluorodeoxyuridine or streptomycin (Wurtz, Sears, Robert, Shepherd, Gillham and Boynton, 1979; Heizmann, Hussein, Nicholas and Nigon, 1982).

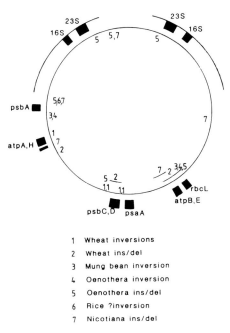

1   Wheat inversions
2   Wheat ins/del
3   Mung bean inversion
4   Oenothera inversion
5   Oenothera ins/del
6   Rice ?inversion
7   Nicotiana ins/del

Fig. 1 - Sites of chromosomal mutations in chloroplast DNA.
The map shows the regions of an ancestral dicot-like chloroplast DNA molecule which have been involved in rearrangements during the evolution of monocots and some dicots (Palmer and Thompson, 1982; Howe *et al.*, 1983; Hirai *et al.*, 1985; Herrmann *et al.*, 1983; Gordon *et al.*, 1982; Kung *et al.*, 1982; Salts *et al.*, 1984; Ogihara and Tsunewaki, 1982; Bowman *et al.*, 1983; Quigley and Weil, 1985).

MOLECULAR ANALYSIS OF AN INVERSION

There is therefore good evidence for the occurrence in the chloroplast of
recombination and related processes. However, little is known about the
mechanism. The end points of an evolutionary rearrangement have recently been
analysed in our laboratory. This is the large inversion in the evolution
of the monocots (which has brought about an inversion of the outermost
points, labelled "l" in Fig. 1 (Howe, 1985a). This particular rearrangement
is likely to be helpful to study, since it is found in only a selection of
monocotyledons, and therefore has probably occurred relatively recently in
evolution. There is therefore a good chance that any sequences(s) which
caused the inversion will still be discernible. The location of the endpoints was first indicated in experiments to map the gene for the alpha
subunit of ATP synthase (Howe et al., 1983). Subsequently DNA sequence
analysis has been carried out, and the organisation of the regions is
displayed in the upper half of Fig.2. The lower half of the figure indicates

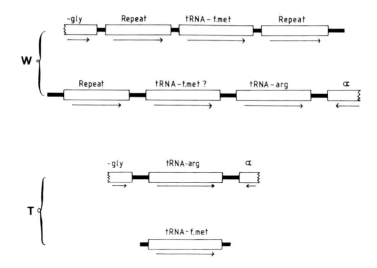

Fig. 2 - Comparison of gene organisation at the inversion endpoints
          in wheat (upper two lines) with tobacco (lower). The
          polarity of the repeats is arbitrarily defined. Genes
          for tRNA's, a possible pseudogene, and part of the gene
          for the alpha subunit for ATP synthase are indicated
          (Howe, 1985; Ohme, et al.,1984; Deno and Sugiura, 1984).

the organisation in tobacco (Ohme, Kamosashira, Shinozoki and Sugiura, 1984; Deno and Sugiura, 1984), which is believed to represent the initial dicot-like form. In wheat, a series of short (*ca.* 70bp) repeats are found, and the inversion could be accounted for by recombination across these. In addition there appears to have been a duplication of the f-met tRNA gene since there are now two copies, one of which is imperfect and may be a pseudogene. Subsequent to the first inversion at least two others have happened (Howe, 1985; Quigley and Weil, 1985) so that the repeats are now present in a direct configuration, as is shown in Fig. 3. Recombination across these would therefore now lead to a deletion. It can be seen from Fig. 2, that the repeats are not found in tobacco, so they may have arisen shortly before the inversion took place. DNA hybridisation experiments (Howe, unpublished results) suggest that the repeats are present also in maize and barley (which also show the inversion), but not in *Spirodela* (which does not). Thus it may be that recombination is simply dependent upon sequence homology. One would therefore expect that the other inversions in Fig. 3 would be associated with repeated sequences, and experiments are in progress to test this. However, part of the region involved is also subject to a high rate of insertion/deletion, and this may have destroyed the sequences in question.

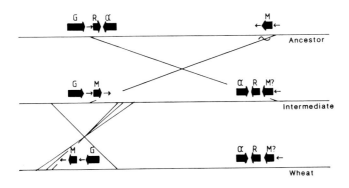

Fig. 3 - Multi-stage derivation of chloroplast DNA in wheat from an ancestral form resembling that found in some dicots. The genes marked are as in Fig. 2. (Howe, 1985; Quigley and Weil, 1985).

Certainly there do not appear to be further copies of the 70bp repeat here, or in the large inverted repeats (Howe, unpublished results).

An intriguing observation is that the 70bp repeat contains a sequence which is remarkably similar to the bacteriophage lambda attachment site. Of the 15bp of the *att* site core, 9 or 10bp are identical in the repeats. This sequence, when cloned in *E. coli*, can in fact function as a lambda attachment site (Howe, 1985a). This should not necessarily be taken to imply a similar function in the chloroplast (i.e. a recognition site for site-specific recombination), but it remains a possibility, and it is tempting to note the wide conservation of recognition sites for certain other site-specific recombination systems (reviewed in Watson, 1985). Furthermore, it may be possible to apply this observation to chloroplast genetic manipulation where, as with other systems, it will be desirable to be able to target foreign DNA to specific sites (De Block, Schell and Van Montagu, 1985; Howe, 1985b; Smithies, Gregg, Boggs, Koralewski and Kucherlapati, 1985).

are in close proximity to tRNA genes. (Howe, 1985a; Quigley and Weil, 1985). While there are, of course, over 30 tRNA genes in wheat ct-DNA (Mubumbila, Bowman, Droog, Dyer, Kuntz and Weil, 1985), they are by no means evenly distributed around the genome. It is possible that some feature of the structure, sequence or transcription of these regions may act to stimulate recombination.

CONCLUSIONS

A number of regions of the chloroplast chromosome appear to be particularly active in promoting recombination and insertion/deletion mutations. In at least one case, they appear to be associated with repeated sequences. These are short, and so pairing between them is likely to be rare by comparison to pairing between the large inverted repeats. Thus recombination across them is likely to be much rarer. It is also possible that recombination may include an element of sequence specificity. There is a need to analyse more sites of ct-DNA rearrangement in a range of species to see what common patterns of sequence or structure emerge. At the same time, it will be important to identify the proteins which are involved, how they work, and how they are regulated.

ACKNOWLEDGEMENT

I am grateful to the Nuffield Foundation for financial support.

REFERENCES

Bohnert, H.J. and Loffelhardt, W. (1982). Cyanelle DNA from *Cyanophora paradoxa* exists in two forms due to intramolecular recombination *FEBS Lett.* 150 403.

Bowman, C.M., Bonnard, G. and Dyer, T.A. (1983). Chloroplast DNA variation between species of *Triticum* and *Aegilops*. Location of the variation on the chloroplast genome and its relevance to the inheritance and classification of the cytoplasm. *Theor. Appl. Genet.* 65 247 - 262.

Gillham, N.W. (1978). Organelle Heredity. Raven Press, New York.

Day, A. and Ellis, T.H.N. (1985). Chloroplast DNA deletions associated with wheat plants regenerated from pollen; possible basis for maternal inheritance of chloroplasts. *Cell 39* 359 - 368.

De Block, M., Schell, J, and Van Montagu, M. (1985). Chloroplast transformation by *Agrobacterium tumefaciens*. *EMBO J.* 4 1367 - 1372.

Deno, H. and Sugiura, M. (1984). Chloroplast tRNA^{Gly} gene contain a long intron in the D stem. Nucleotide sequences of tobacco chloroplast genes for tRNA^{Gly} (UCC) and the tRNA^{Arg} (UCU). *Proc. Natl Acad. Sci. USA* 81 405 - 408.

Fluhr, R. and Edelman, M. (1981). Conservation of sequence arrangement among higher plant chloroplast DNAs: molecular cross hybridisation among the Solanaceae and between *Nicotiana* and *Spinacia*. *Nucleic Acids Res.* 9 6841 - 6853.

Gordon, K.H.J., Crouse, E.J., Bohnert, H.J. and Herrmann, R.G. (1982). Physical mapping of differences in chloroplast DNA of the five wild-type plastomes in *Oenothera* subsection *Euoenothera Theor. Appl. Genet.* 61 373 - 384.

de Heij, H.T., Lustig, H., Moeskops, D.J.M., Bovenberg, W.A., Bisanz, C. and Groot, G.S.P. (1983). Chloroplast DNAs of *Spinacia, Petunia* and *Spirodela* have a similar gene organisation. *Curr. Genet.* 7 1 - 6.

Heizmann, P., Hussein, Y., Nicolas, P. and Nigon, V. (1982). Modifications of chloroplast DNA during streptomycin-induced mutagenesis in *Euglena gracilis*. *Curr. Genet.* 5 9 - 15.

Herrmann, R.G., Westhoff, P., Alt, J., Winter, P., Tittgen, J., Bisanz, C., Sears, B.B., Nelson, N., Hurt, E., Hauska, G., Viebrock, A. and Sebald, W. (1983). Identification and characterization of genes for polypeptides of the thylakoid membrane in Structure and Function of Plant Genomes (eds. Ciferri, O. and Dure, L. pp 143 - 153) Plenum Press, New York.

Hirai, A., Ishibashi, T., Morikami, A., Iwatsuki, N., Shinozaki, K. and Sugiura, M., Rice chloroplast DNA: a physical map and the location of the genes for the large subunit of ribulose-1,5-bisphosphate carboxylase and the 32kD photosystem II reaction center protein. *Theor. Appl. Genet.* 70 117 - 122. (1985)

Howe, C.J., Bowman, C.M., Dyer, T.A. and Gray, J.C. (1983). The genes for the alpha and proton-translocating subunits of wheat chloroplast ATP synthase are close together on the same strand of chloroplast DNA *Mol. Gen. Genet.* 190 51 - 55.

Howe, C.J. (1985). The endpoints of an inversion in wheat chloroplast DNA are associated with short repeated sequences showing homology to *att*-lambda. *Curr. Genet.* 10 in press.

Howe, C.J. (1986). Chloroplast transformation by *Agrobacterium tumefaciens*. *Trends in Genetics 1* 217 - 218.

Kolodner, R. and Tewari, K.K. (1979). Inverted repeats in chloroplast DNA from higher plants. *Proc. Natl. Acad. Sci. USA 76* 41 - 45.

Kung, S.D., Zhu, Y.S. and Shen, G.F. (1982) *Nicotiana* chloroplast genome III. Chloroplast DNA evolution. *Theor. Appl. Genet.* 61 73 - 79.

Mubumbila, M., Bowman, C.M., Droog, F., Dyer, T.A., Kuntz, M. and Weil, J.H. (1985). Chloroplast transfer RNAs and tRNA genes of wheat. *Plant Mol. Biol.* 4 315 - 320.

Ogihara, Y. and Tsunewaki, K. (1982). Molecular basis of the genetic diversity of the cytoplasm in *Triticum* and *Aegilops* I. Diversity of the chloroplast genome and its lineage revealed by the restriction pattern of chloroplast DNAs. *Jpn. J. Genet.* 57 371 - 396.

Ohme, M., Kamogashira, T., Shinozoki, K. and Sugiura, M. (1984). Locations and sequences of tobacco chloroplast genes for tRNA^{Pro} (UGG), tRNA^{Trp}, tRNA^{f-Met} and tRNA^{Gly} (GCC): the tRNA^{Gly} contains only two base pairs in the D-stem. *Nucleic Acids Res.* 12 6741 - 6749.

Oliver, R.P. (1984). Location of the genes for cytochrome f, subunit IV of the b$_6$/f complex, the   subunit of CF$_1$ ATP synthase and subunit III of the CF$_0$ ATP synthase on the barley chloroplast chromosome. *Carlsberg Res. Commun.* 49 555 - 557.

Palmer, J.D. and Thompson, W.F. (1982). Chloroplast DNA rearrangements are more frequent when a large inverted repeat sequence is lost. *Cell 29* 537 - 550.

Palmer, J.D. (1983). Chloroplast DNA exists in two orientations. *Nature 301* 92 - 93.

Palmer, J.D., Osorio, B., Watson, J.C., Edwards, H., Dodd, J. and Thompson, W.F. (1984). Evolutionary aspects of chloroplast genome expression and organization in Biosynthesis of the photosynthetic apparatus: Molecular biology, development and regulation. (eds. Thornber, J.P., Staehelin, A, and Hallick, R.B.) pp273 - 283 A.R. Liss, New York.

Quigley, F. and Weil, J.H. (1985). Organization and sequence of five tRNA genes and of an unassigned reading frame in the wheat chloroplast genome: evidence for gene rearrangements during the evolution of chloroplast genomes. *Curr. Genet. 9* 495 - 503.

Sager, R. and Ramanis, Z. (1970). A genetic map of non-Mendelian genes in *Chlamydomonas.Proc. Natl. Acad. Sci. USA 65* 593 - 600.

Salts, Y., Herrmann, R.G., Peleg, N., Lavi, U., Izhar, S., Frankel, R. and Beckmann, J.S. (1984). Physical mapping of plastid DNA variation among eleven *Nicotiana* species. *Theor. Appl. Genet. 69* 1 - 14.

Smithies, O., Gregg, R.G., Boggs, S.S., Koralewski, M.A. and Kucherlapati, R.S. (1985). Insertion of DNA sequences into the human chromosomal -globin locus by homologous recombination. *Nature 317* 230 - 234.

Stahl, F.W., Crasemann, J.M. and Stahl, M.M. (1975). Rec-mediated recombination hot-spot activity in bacteriophage lambda. *J. Mol. Biol. 94* 203 - 212.

Watson, M. (1985). Prokaryote invertible DNA systems are highly conserved. *Trends Biochem. Sci. 9* 82 - 83.

Wurtz, E.A., Sears, B.B., Rabert, D.K., Shepherd, H.S., Gillham, N.W. and Boynton, J.E. (1979). A specific increase in chloroplast gene mutations following growth of *Chlamydomonas* in 5-fluorodeoxyuridine. *Mol. Gen. Genet. 170* 235 - 242.

Molecular analysis of the
mitochondrial genomes of plants

8 Molecular analysis of hypervariability in the mitochondrial genome of tissue cultured cells of maize and sorghum

P. S. Chourey, R. E. Lloyd, D. Z. Sharpe and N. R. Isola

ABSTRACT

Our interests lie in the understanding of the molecular basis of tissue culture variability. Mitochondrial (mt-DNA) is of ideal size and nature to conduct such analyses. Several independently derived cell cultures of maize and sorghum have been examined. A high level of variability is seen specifically among protoclone cultures. Although a large number of restriction fragments are conserved, a certain proportion of the genome seems highly variable. The use of cloned mt-DNA fragments as hybridization probes reveals that repeated DNA regions are specifically associated with a high capacity for loss and/or rearrangement in the genome.

INTRODUCTION

Genetic variability in tissue cultured cells and regenerated plants is extensively documented. Larkin and Scowcroft (1981) provide an excellent overview concerning such variability in several crop plant species. The evidence available so far suggests that spontaneous variability is the rule rather than the exception in cultured cells and regenerated plants. Nondiploid and asexually propagated plants show a higher level of variability than diploid and sexually propagated species. Also, tissue cultured cells show a much higher level of variability than the corresponding regenerated plants. Limited studies have been done on cell culture or plant regenerants derived from protoplasts. Probably the most detailed studies are reported by Shepard, Bidney and Shahin (1980) in potato. Clonal plant populations regenerated from single leaf protoplasts, designated as protoclones, displayed a high frequency of variation for disease resistance and several horticultural characteristics. Lorz and Scowcroft (1983) analyzed variability among plants and their progeny

177

regenerated from *Nicotiana tabacum* protoplasts which are heterozygous for a single locus affecting plant pigmentation. Somatic alterations were far higher in the leaf cells of regenerated plants after prolonged cell culture as compared to seed derived plants. Essentially similar results have been reported by Prat (1983). Barbier and Dulieu (1983 have concluded that such variations are induced during the early stages of protoplast regeneration.

Although extensive documentation of tissue culture variability is available in numerous plant species, most of the work has been done at the morphological, cytological and cytogenetic level. Molecular analyses are limited. Because restriction endonucleases cleave DNA molecules at only specific recognition sites usually four, five or six nucleotides in length, they allow recognition of specific alterations at the DNA level. Nucleotide substitutions that generate or destroy the restriction sites will alter the pattern of cleavage products. Addition, loss or rearrangement of DNA sequences can also be analyzed. The mitochondrial genome, smaller in size than the nuclear genome, has been of particular interest in these analyses. Gengenbach, Connelly, Pring and Conde (1981), Kemble, Flavell and Brettell (1982), McNay, Chourey and Pring (1984) and Kemble and Shepard (1984) have all reported a high level of molecular variability in regenerants from tissue culture derived cells and calli.

The main objective of our research is to characterize and understand the molecular nature of variability induced in tissue cultured cells rather than regenerated plants. We believe that a substantial proportion of variability is lost in the latter due to selection pressure against cells which are not tolerated by functional constraints imposed by development and differentiation. We have given particular attention to the mitochondrial (mt) genome of maize and sorghum for several reasons: Maize mt genome is one of the most extensively analyzed among higher plant mt genomes (Lonsdale, Lodge and Fauron 1984; Levings, 1983). Nearly all recombinant mt-DNA clones from maize share homology with sorghum genome. Consequently many well characterized DNA clones for specific regions such as protein and RNA coding genes, transposable elements, repeated DNA regions etc., are available for both corn and sorghum genomes. An equally important attribute is our tissue culture capability allowing protoplast culture and callus regeneration for both corn (Chourey and Zurawski, 1981) and sorghum

(Chourey and Sharpe, 1985); plant regeneration from these cultures is, however, not yet possible.

The initial study of the NK 3000 cultivar of sorghum included a comparative analysis of mt DNA from six randomly selected protoclone cell cultures, the parental and a new cell suspension culture and coleoptile tissue. The following observations are noteworthy (Wilson, Chourey and Sharpe, 1984):

1. The *EcoR1* digested mt-DNA fragments were visualized both by UV fluorescence of ethidium bromide stained gels and by membrane hybridisation. Each of the above samples displayed a unique pattern. Although a large number of fragments are conserved (i.e., not variable among the samples), a certain proportion of the genome seems to be variable. Two of the six protoclones show a complete loss of the largest fragment.

2. This loss was further confirmed by Southern blot (S.b.) hybridisation analysis using the largest *EcoR1* fragment as the hybridisation probe. Each sample displayed mutual hybridisation of more than one fragment; protoclones showed the largest number of such homologous fragments. This hybridisation pattern was indicative of the involvement of repetitive DNA or transposable elements in accounting for this variability. The protoclonal variability for the largest *EcoR1* fragment appears to be independent of cell cycle specific events in the parental cell line.

3. This fragment has now been cloned into lambda phage DNA (Zack and Chourey, 1985). The recombinant phage, designated as S.b. 14.7 has been used as the hybridisation probe on mt-DNA digests from the above samples and similar results have been obtained. Cell clone cultures were also analyzed and hybridisation patterns similar to the parental cell suspension cultures are seen (Wilson *et al.*, manuscript in preparation). The high levels of variability are limited to the protoclone cultures.

4. Coleoptile mt-DNAs of five cultivars of sorghum derived from diverse genetic backgrounds show a high level of conservation for the fragments hybridising to this recombinant clone (Zack and Chourey, 1985).

MATERIALS AND METHODS

Maintenance of cell suspension cultures, isolation of protoplasts, subsequent regeneration and growth leading to callus formation was done according to Chourey and Zurawski (1981) and Chourey and Sharpe (1985) for maize and sorghum respectively. Protoplast-derived calli were transferred back to liquid medium to obtain protoclone cell suspension cultures. To obtain cell-clone cultures, small cell aggregates from cell suspensions were plated at low densities so that individual colonies were separable with the naked eye. Each individual callus was grown separately for an additional 3-4 weeks and then transferred to liquid medium.

Etiolated coleoptile tissue from 5-7 day old seedlings, grown in vermiculite, were harvested for mt-DNA isolations. Tissue cultured cells were harvested during the logrithmic phase of growth. All mt-DNA isolations were performed according to Wilson and Chourey (1984). Restriction digests of mt-DNAs were done according to manufacturers specifications. Horizontal agarose gel electrophoresis was conducted in o.8% agarose gel using TPE buffer (0.03M Tris-HCl, 0.036M NaH_2PO_4, 0.001M Na_2EDTA pH 7.8) at room temperature for 15 h at 3V/cm. Gels were stained with ethidium bromide (0.5 μg/ml final concentration) for one hour, destained with water and photographed under UV light. The mt-DNA samples with evidence of complete restriction digestion were transferred from agarose gels to nitrocellulose as described by Southern (1975).

DNA was labelled by nick translation using (α-^{32}P) dCTP as described by Rigby, Diekman, Rhodes and Berg (1977). Unincorporated deoxynucleotide triphosphates were removed by Sepadex G50 column chromatography. Prehybridisation, hybridisation, washing and autoradiography of Southern blots were done as described previously (McNay et al., 1984).

For cytological analyses, cells at logarithmic phase in suspension cultures were pretreated with 0.004M 8-hydroxyquinoline for 4 h at 10°C and fixed in acetic alcohol overnight. Chromosome counts were determined from macerated tissue stained with Carbol fuchsin (Kao, 1975).

RESULTS AND DISCUSSION

A) Maize

1) The 14.7 kb region
The observed hypervariability in sorghum protoclones, as visualised by hybridisation with the λS.b.14.7 clone, led us to examine maize cultures relative to this region of the sorghum mt-DNA. Southern blot hybridisation blots of Black Mexican Sweet (BMS) maize cell suspension, two protoclone cell cultures and the BMS coleoptile mt-DNA cleaved with EcoRl and probed

with S.b.14.7 clone are shown in Fig. 1, panel A. Two major hybridising fragments are seen in all the samples. However, the parental cell suspension and the second protoclone of independent origin (PC 1) show two additional smaller sized fragments (A, lanes 2 and 3). Although no differences are seen in BamHl digest analysis (Fig. 1, panel B), the *Hind* III pattern resolves these samples again into two classes (Panel C) similar to that seen by the *EcoRl* enzyme. Additional protoclone cell cultures are being isolated to ascertain the level of hypervariability in maize samples.

A BMS mt-DNA library prepared by A.G. Smith in D.R. Pring's laboratory has been screened for λS.b. hybridising clones. Six cosmid clones have

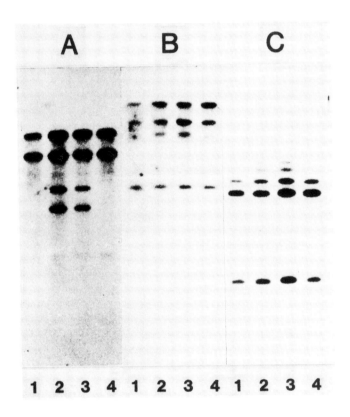

Fig. 1 – Southern hybridisation analysis of BMS maize mt-DNA
samples digested with *EcoR*I (A), *Bam*HI (B) or *Hind*III
(c) and hybridised with S.b. 14.7 probe. Samples
include: 1) coleoptile 2) cell suspension, 3) and 4)
protoclone 1 and 2 respectively.

been isolated by dot blot analysis; however, none show the simultaneous presence of both *EcoRl* fragments by Southern hybridisation analysis (Zack and Chourey, unpublished data). This indicates that the two regions are spatially separable on the 'master' molecule (Lonsdale *et al.*, 1984) or that they are on two different molecules. Physical mapping of these regions may provide a better understanding of genetic mechanisms leading to tissue culture variability.

2) Interorganellar Region

Stern and Lonsdale (1982) have identified maize mt-DNA sequences which are homologous to parts of the maize chloroplast (ct) genome and are believed to have arisen due to random interorganellar integration during evolution (Lonsdale, 1985). A 12 kb sequence in a mt-cosmid designated as 8-3H4 (Stern and Lonsdale, 1982) is of special interest. It hybridises to 11 *Bam*Hl ct fragments; 6 of the 11 fragments co-migrate with mt fragments and the largest such fragment is of 4.8 kb size. Because ct genomes are highly conserved in various plant species and also in tissue culture regenerated plants (Kemble and Shepard, 1984), it was of interest to examine the stability of ct sequences present in mitochondria. Hybridisation analysis with 8-3H4 DNA as the probe on a *Bam*Hl digest of mt-DNA from various samples is shown in Fig. 2. Of the many restriction fragments seen in this Southern blot, only one fragment of 4.8 kb size, seen in the coleoptile and a protoclone digest (panel B, lanes 1 and 4, respectively) is missing in the cell suspension culture and another protoclone (PC 1) culture (panel B, lanes 2 and 3, respectively). Instead, a new fragment of slightly larger size is seen in these two cultures. Because no other alterations are visualized, we speculate that the entire 4.8 kb region is now re-arranged to this larger fragment.

It is worth noting that the parental BMS cell suspension culture exhibits the initial change from 4.8 kb size to a larger fragment. One of the protoclone cultures (PC 2) derived from this cell suspension, however, shows a reversion to the original 4.8 kb size as seen in the BMS coleoptile. Such a specific change, is reminiscent of precise excision of transposable elements leading to a restoration of the wild type phenotype (Fedoroff, 1983). Whether this fragment has transposable element-like properties or unique recombinogenic sequences is at present not known. A better characterisation

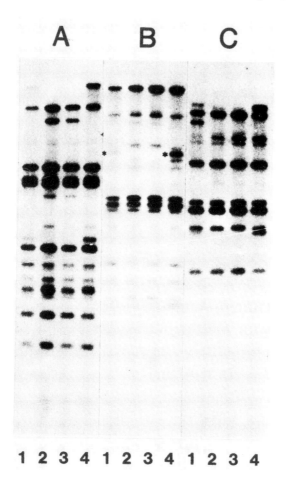

Fig. 2 - Southern hybridisation analysis of BMS maize mt-DNA
samples digested with *Eco*RI (A), *Bam*HI (b) or *Hind*III
(c) and hybridised with cosmid clone 8-3H4 probe.
Samples include: 1) coleoptile 2) cell suspension, 3)
and 4) protoclone 1 and 2 respectively. The asterisks
in B identify the 4.8 kb fragment.

of this region through sequencing of the junction regions will certainly
contribute to a better understanding of the molecular nature of tissue
culture variability. Similar hybridisation analysis with 8-3H4 clone of
*Hind*III digested mt-DNA samples was also performed (Fig. 2, panels A and
C). Considerable variation is seen among the samples and some of it is
presumably related to the 4.8 kb *Bam*Hl fragment.

Mitochondrial genomes

B. Sorghum

1. Cytochrome oxidase regions

The high level of mt-DNA variability in sorghum protoclone cultures (Wilson
et al., 1984; Zack and Chourey, 1985; Wilson *et al.*, manuscript in preparation) has led us to examine regions of the genome corresponding to known
functions such as protein coding genes. The present analysis is, however,
restricted to regions encoding subunit I and II of the *cytochrome oxidase*
(CO) protein. Assuming that these two genes are expressed in tissue cultured
cells and are essential for mitochondrial biogenesis and for normal cell
division processes, stringent functional constraint(s) would presumably
result in the maintenance of nonvariant copies of these genes. Samples
analysed include the same six protoclone cell cultures which were previously
examined (Wilson *et al.*, 1984), six new cell-clone cultures, several cell

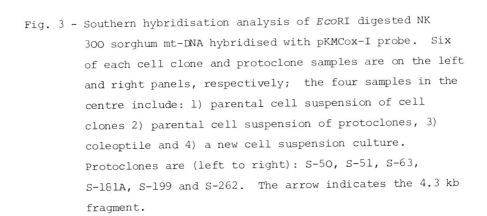

Fig. 3 - Southern hybridisation analysis of *Eco*RI digested NK
300 sorghum mt-DNA hybridised with pKMCox-I probe. Six
of each cell clone and protoclone samples are on the left
and right panels, respectively; the four samples in the
centre include: 1) parental cell suspension of cell
clones 2) parental cell suspension of protoclones, 3)
coleoptile and 4) a new cell suspension culture.
Protoclones are (left to right): S-50, S-51, S-63,
S-181A, S-199 and S-262. The arrow indicates the 4.3 kb
fragment.

suspension cultures which are related to the derived cell-clones and protoclones and the parental coleoptile mt-DNAs. All samples are derived from the NK 300 cultivar. Southern blots of *Eco*Rl digested samples are hybridised with nick translated plasmid DNAs corresponding to *COI* and *COII* clones and are shown in Figs. 3 and 4 respectively.

Hybridisation analysis with the *COI* clone which contains the entire *COI* gene on a 4.3 kb *Eco*Rl sorghum insert (Bailey-Serres, Hanson, Fox and Leaver, 1984) reveals a single major fragment of the same size in all the samples (Fig. 3). No variations are seen. A faint band of smaller molecular size is also observed but significance of this hybridisation is unclear. Similar analysis with the *COII* clone yields an intensely hybridising fragment of 1.3 kb corresponding to a part of the *COII* insert in the plasmid

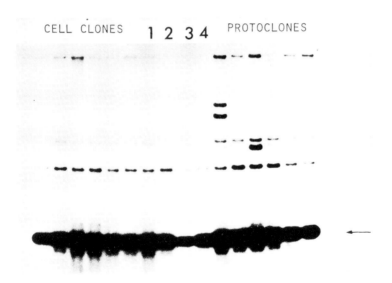

Fig. 4 - Southern hybridisation analysis of *Eco*RI digested NK
300 sorghum mt-DNA hybridised with pk9ECoxII probe. Six
of each cell clone and protoclone samples are on the left
and right panels respectively; the four samples in the
centre include: 1) parental cell suspension of cell clones
2) parental cell suspension of protoclones 3) coleoptile
and 4) a new cell suspension culture. Protoclones are
(left to right): S-50, S-63, S-181A, S-199 and S-262.
The arrow idicates the 1.3 kb fragment.

in all the samples. In addition, at least three non-variant weakly
hybridising bands of higher molecular weight are also seen. The most
critical observation on this hybridisation membrane is that protoclone
S-50 (Fig. 4) shows two intensely hybridising fragments not seen in any
other sample. We believe this alteration is specific to the 1.3 kb insert
which contains exon-1 and the partial region corresponding to the intron of
the *COII* gene. The *COII* gene has been intensively examined in maize
(Fox and Leaver, 1981), wheat (Bonen, Boer and Gray, 1984) and rice (Kao,
Moon and Wu, 1984). The coding region of this gene in both wheat and rice
is 99% or more similar to the maize gene. An intron of variable size is
centrally located in each case; the intron size variation in wheat and rice
relative to maize is due to an insertion which exhibits transposable element-
like structure (Bonen *et al.*, 1984; Kao *et al.*, 1984). Although the sorghum *COII* gene remains to be analyzed, it is safe to assume that it also
has an intron as seen in other monocots. Further analyses on the nature
of the unique hybridising fragments in protoclone S-50 will provide significant information. It is obvious that the *COII* hybridising regions are
rearranged on larger fragments. It is possible that the intron region of
the gene on these fragments is expanded, as in wheat and rice during evolution.
If a full length *COII* gene on these fragments is demonstrated, it will be
an important variant for gene expression analyses.

Finally, it is important to note that both *COI* and *COII* genes are
conserved during tissue culture. Additional hybridising fragments are
restricted to *COII* and not the *COI* gene. It is tempting to speculate that
the lack of variability in *COI* may be related to the lack of an intron in
this gene. However, an alternate possibility such as the presence or
absence of repeated DNA elements in the adjoining region of the gene is
not ruled out.

2. Nuclear ploidy analysis

Nuclear influence on the maintenance of mt-DNA organisation in mammalian
cells (DeFrancesco, Attardi and Croce, 1980) and in plants (Laughnan,
Gobey-Laughnan and Carlson, 1981; Borck and Walbot, 1982) has been
demonstrated. The high level of mt-DNA variability in protoclone cultures
prompted us to examine the range of chromosome number variability in six
protoclone cultures against the parental cell culture. Polyploidy,

aneuploidy and several other cytological aberrations are a common feature of long term propagation of cells in a tissue culture environment. It is conceivable that protoclone cultures may have an even higher level of such variability than the parental cell suspension culture. The doubling of chromosome number may arise as a consequence of either interruptions in cell division processes or spontaneous protoplast fusion (homokaryotic) during any one of the several steps in protoplast isolation and culture. Endo-reduplication, polyploidisation and aneuploidy during the initial stages of protoplast growth and division in potato shoot protoplasts appear to be common events (Sree Ramulu, Dijkhuis, Roest, Bokelmann and De Groot, 1984; Carlberg, Glimelius and Eriksson, 1984).

Table I Analysis of mean and range of variation in chromosome number in cell suspension and protoclone cultures

NK 300 Cell Suspension	Protoclone			Cell Suspension		
	S50	S5L	S63	S181	S199	S262
Mean+S.E. 76.5+3.5	64.0+2.4	89.0+1.8	84.2+4.4	67.2+2.2	65.2+2.3	58.7+2.5
(n=20 cells)						
Range 40-98	38-78	78-104	40-122	50-84	40-84	36-74

NK 300 root tip counts 2n=20

The chromosome number analysis of sorghum cultures is shown in Table I. A total of 20 cells in each culture were examined. No diploid cell (2n=20) was detected in any of the seven cultures. More importantly, the range and mean chromosome number in the six protoclone cultures is not significantly different from those seen in the parental cell suspension culture. This is surprising and indicates that cytological variability induced at the initial stages of the protoplast culture process either tends not to prevail in the end or that the same type of events (e.g., endo-reduplication, polyploidisation etc.) is also encountered in the parental cell suspension. However, the nuclear role on the mt-DNA variations cannot be ruled out from this limited cytological analysis which provides no information on various qualitative changes in the chromosomes. In addition, very little is known concerning the coordination between nuclear DNA and organellar DNA replications. The protoplast isolation process impairs or delays the cell division process in an obligatory fashion due to the lack of a cell wall. It will be surprising if this trauma alone does not lead to organellar DNA variations.

CONCLUSIONS

The main aim of our studies is to understand better the molecular nature of genetic variability in tissue cultured cells. We are using the mt genome of maize and sorghum as a model system to analyse this problem. The approach includes the use of well characterized cloned DNA fragments as hybridisation probes to monitor variability in specific regions of the cellular genome. Evidence available so far from this as well as other related on-going studies suggests that the variability is non-random in nature. Variable regions appear to be associated with repeated DNA elements and/or transposable elements. Further analyses are needed, however, to make a conclusive judgement. In light of the variability seen here in proto-clone cultures, we are in full agreement with Kemble and Shepard (1984) that considerable caution is needed concerning the interpretation of mt-DNA recombination events in somatic hybrids.

ACKNOWLEDGEMENTS

We are grateful to Drs. C.J. Leaver and J. Bailey-Serres for sharing sorghum clones corresponding to *COI* and *COII* region prior to publication of their data and to Dr. D.M. Lonsdale for cosmid clone 8-3H4 used in the studies described here. The senior author thanks Drs. D. R. Pring and D.M. Lonsdale for many interesting discussions concerning various aspects of maize and sorghum mitochondrial genomes. This cooperative investigation was between the U.S. Department of Agriculture, ARS and IFAS and was supported in part by the U.S. Department of Agriculture, Competitive Grants Office, Grant No.85-CRCR-1-1545.

REFERENCES

Bailey-Serres, J., Hanson, D.K., Fox, T.D. and Leaver, C.J. (1984) Expression of a variant gene for ctyochrome *c* oxidase subunit I in a cytoplasmic male sterile sorghum. *EMBO Workshop: Plant Mitochondrial DNA*, Scotland, P.31.

Barbier, M. and Dulieu, H. (1983). Early occurrences of genetic variants in protoplast cultures. *Plant Sci. Lett., 29:* 201 - 206.

Bonen, L., Boer, P.H. and Gray, M.W. (1984). The wheat cytochrome oxidase subunit II an intron and three radical amino acid changes relative to maize. *The EMBO J., 3:* 2531 - 2436.

Borc, K.S. and Walbot, V. (1982). Comparisons of the restriction endonu-clease patterns of mitochondrial DNA from normal and male-sterile cytoplasms of *Zea mays* L. *Genetics, 102:* 109 - 128.

Carlberg, I., Glimelius, K. and Eriksson, T. (1984). Nuclear DNA content during the initiation of callus formation from isolated protoplasts of *Solanum tuberosum* L. *Plant Sci. Lett., 35:* 225 - 230.

Chourey, P.S. and Sharpe, D.Z. (1985). Callus formation from protoplasts of *Sorghum* cell suspension cultures. *Plant Sci., 39:* 171 - 175.

Chourey, P.S. and Zurawski, D.B. (1981). Callus formation from protoplasts of a maize cell culture. *Theor. Appl. Genet., 59:* 341 - 344.

DeFrancesco, L., Attardi, G. and Croce, C.M. (1980). Uniparental propaga-tion of mitochondrial DNA in mouse-human hybrids. *Proc. Natl. Acad. Sci. USA, 77:* 4079 - 4083.

Fedoroff, N.V. (1983). Controlling elements in maize. In: *Mobile Genetic Elements.* (ed. Shapiro, J.A.) pp. 1 - 63, Academic Press.

Fox, T.D. and Leaver, C.J. (1981). The *Zea mays* mitochondrial gene coding cytochrome oxidase II has an interesting sequence and does not contain TGA co-ons. *Cell, 26:* 315 - 323.

Gengenbach, B.G., Connelly, J.A., Pring, D.R. and Conde, M.F. (1981). Mitochondrial DNA variation in maize plants regenerated during tissue culture selection. *Theor. Appl. Genet., 59:* 161 - 167.

Kao, K.N. (1975). A nuclear staining method for plant protoplasts. In: *Plant Tissue Culture Methods.* (eds. Gamborg, O.L. and Wetter, L.R.) pp. 60 - 62, National Research council of Canada.

Mitochondrial genomes

Kao, T., Moon, E. and Wu, R. (1984). Cytoplasmic oxidase subunit II gene of rice has an insertion sequence within the intron. *Nucleic Acids Res.,* *12:* 7305 - 7315.

Kemble, R.J. and Shepard, J.F. (1984). Cytoplasmic DNA variation in a potato protoclonal population. *Theor. Appl. Genet., 69:* 211 - 216.

Kemble, R.J., Flavell, R.B. and Brettell, R.I.S. (1982). Mitochondrial DNA analysis of fertile and sterile maize plants derived from tissue culture with the Texas male sterile cytoplasm. *Theor. Appl. Genet.,* *62:* 213 - 217.

Larkin, P.J. and Scowcroft, W.R. (1981). Somaclonal variation - a novel source of variability from cell cultures for plant improvement. *Theor. Appl. Genet., 60:* 197 - 214.

Laughnan, J.R., Gabay-Laughnan, S. and Carlson, J.E. (1981). Characteristics of *cms-S* reversion to male-fertility in maize. *Stadler Genet. Symp., 13:* **93** - 114.

Levings, C.S. III (1983). The plant mitochondrial genome and its nutants. *Cell, 32:* 659 - 661.

Lonsdale, D.M. (1985). Chloroplast DNA sequences in the mitochondrial genome of maize. In: *Molecular Form and Function of the PlantGenome.* (eds. van Vloten-Doting, L., Groot, G.S.P. and Hall, T.C.) pp. 421 - 428, Plenum.

Lonsdale, D.H., Hodge, T.P. and Fauron, C.M.-R. (1984). The physical map and organisation of the mitochondrial genome from the fertile cytoplasm of maize. *Nucleic Acids Res., 12:* 9249 - 9261.

Lorz, H. and Scowcroft, W.R. (1983). Variability among plants and their progeny regenerated from protoplasts of *Su/Su* heterozygotes of *Nicotiana tabacum. Theor. Appl. Genet., 66:* 67 - 75.

McNay, J.W., Chourey, P.S. and Pring, D.R. (1984). Molecular analysis of genomic stability of mitochondrial DNA in tissue cultured cells of maize. *Theor. Appl. Genet., 67:* 433 - 437.

Prat, D. (1983). Genetic variability induced in *Nicotiana sylvestris* by protoplast culture. *Theor. Appl. Genet., 64:* 223 - 230.

Rigby, P.W.J., Dieckman, M., Rhodes, C. and Berg, P. (1977). Labelling deoxyribonucleic acid to high specific activity *in vitro* by nick-translation with polymerase I. *J. Mol. Biol., 113:* 237 - 251.

Shepard, J.F., Bidney, D. and Shahin, E. (1980). Potato protoplasts in crop improvement. *Science, 208:* 17 - 24.

Southern, E.M. (1975). Detection of specific sequences among DNA fragments separated by gel electrophoresis. *J. Mol. Biol., 98:* 503 - 517.

Sree Ramulu, K., Dijkhuis, P., Roest, S., Bokelmann, G.S. and DeGroot, B. (1984). Early occurrence of genetic instability in protoplast cultures of potato. *Plant Sci. Lett., 36:* 79 - 86.

Stern, D.B. and Lonsdale, D.M. (1982). Mitochondrial and chloroplast genomes of maize have a 12-kilobase DNA sequence in common. *Nature 299:* 698 - 702.

Wilson, A.J. and Chourey, P.S. (1984). A rapid inexpensive method for the isolation of restrictable mitochondrial DNA from various plant sources. *Plant Cell Reports, 3:* 237 - 239.

Wilson, A.J., Chourey, P.S. and Sharpe, D.Z. (1984). Protoclones of *Sorghum bicolor* with unusually high mitochondrial variation. In: *Symp. Tissue Culture in Forestry and Agriculture.* (eds. Henke, R.R., Hughes, K.W., Constantin, M.J. and Hollander, A.) pp. 368 - 369. Plenum.

Zack, C.D. and Chourey, P.S. (1985). Molecular characterization of a region of mitochondrial DNA which is hypervariable in cultured cells of *Sorghum bicolor* cv NK 300. *First Int'l. Congress of Plant Molec. Biol.* (in press).

9 Analysis of the protoplast fusion-induced molecular events responsible for the mitochondrial DNA polymorphism in the rapeseed *Brassica napus*

F. Vedel, P. Chetrit, C. Mathieu, G. Pelletier and C. Primard

ABSTRACT

Analysis of the cytoplasmic genomes of six rapeseed somatic hybrid plants strongly suggest that cytoplasmic male sterility (cms) is encoded in mitochondria. The cms phenotype segregates independently of parental plastid DNA type. The mt-DNA restriction patterns differ from each other and from those of the partners of fusion. The new DNA fragments in the cybrid mitochondrial DNAs result from reciprocal recombination between the two parental mitochondrial genomes. Protoplast fusion-induced recombination and intragenomic natural recombination seem to involve the same specific sites in the mitochondrial genomes in *Cruciferae*. Mapping experiments are in progress to localize the mitochondrial sterility factors.

INTRODUCTION

In higher plants, cytoplasmic male sterility (cms) is a maternally inherited trait which prevents the production of functional pollen but does not affect female fertility (Duvick 1965). This trait is widely used in the commercial production of Fl hybrid seed varieties by preventing self pollination in numerous crop species. A cms character is qualitatively defined by a specific genetic system of fertility restoration. The use of cms lines as female parents in the commercial production of Fl hybrids was facilitated by the discovery of fertile lines that carry specific dominant nuclear genes which restore fertility by suppressing the male sterile phenotype. This results in the Fl hybrid progeny being male fertile, even though they contain the male sterile cytoplasm. There is some biochemical evidence that the mitochondrial genome rather than the chloroplast genome is the carrier of male sterility factors in the cytoplasmic compartment. Namely, analysis of somatic hybrids gives some clues proving that cms is encoded in mitochondria. Recently, Pelletier, Primard, Vedel, Chetrit, Remy, Rousselle and Renard (1983) have regenerated cytoplasmic somatic hybrids of *Brassica napus* combining *Brassica* chloroplasts and a cms trait from *Raphanus sativus*. Analysis of the cytoplasmic DNAs isolated from these cybrids indicates that (1) the cms phenotype segregates independently of parental plastid type. (2) protoplast fusion leads to new mitochondrial genomes that could arise through intermolecular recombination (Chetrit, Matthieu, Veld, Pelletier and Primaval, 1985). These results were in agreement with previous findings on cytoplasmic hybrids from Solanaceous species (Belliard, Pelletier, Vedel and Quetier, 1978; Belliard, Vedel and Pelletier, 1979; Nagy, Terok and Malign, 1981; Galun, Arzee-Gonen, Fluhr, Edelmann and Aviv, 1982; Boeshore, Lifshz, Hanson and Izher, 1983; Fluhr, Aviv, Edelmann and Galun, 1983). The hypothesis of mitochondrial recombination is necessary to explain the physical maps of higher plant mt DNAs (Palmer and Shields, 1984; Chetrit, Matthieu, Muller and Veld, 1984; Lonsdale, Hodge and Fauron, 1984; Falconet, Lejeune, Quetier and Gray, 1984). The mitochondrial genomes of *Brassicae* are the smallest known in higher plants (Lebacq and Vedel, 1981) and appear very valuable in studying mt-DNA organisation and the fate of parental mt-DNA in cybrids.

In this paper, we describe (1) the phenotypes and the cytoplasmic genotypes of several *Brassica* cybrids, (2) the mt-DNA polymorphism of these cybrids, (3) evidence for mitochondrial intergenomic recombination from the hybrid constitution of a novel restriction fragment, (4) homology between mitochondrial sites involved in naturally occurring recombination and mitochondrial sites involved in protoplast fusion induced recombination, and (5) restoration fertility experiments corroborating the molecular analysis of mitochondrial recombination events.

PHENOTYPICAL AND GENOTYPICAL CHARACTERISTICS IN CYTOPLASMIC PARASEXUAL HYBRIDS OF *B. NAPUS*

The different *B. napus* lines used in the fusion experiments have been described by Pelletier *et al.*, (1983). Three cultivars of rapeseed were used as a source of protoplasts:
1) pure line, "Brutor" variety, the nucleus of which is in a homologous *B. napus* cytoplasm leading to the following cytoplasmic characters: normal green leaves, well-developed nectaries (N^+): atrazine susceptibility (Atr^s) and male fertility (F).

2) "C" genotype, a cms O line of *B. napus* obtained by sexual crosses involving a cms Japanese radish variety (Ogura 1968). Plants have 38 chromosomes and have lost all *R. sativus* chromosomes. Their *R. sativus* cytoplasm leads to four cytoplasmic chracters: yellow leaves (chlorophyll deficient) at low temperature, underdeveloped nectaries (N^-), atrazine susceptibility (Atr^s) and male sterility (cms).

3) "Tower" cultivar a line of *B. napus* arising from sexual crosses involving a triazine resistant biotype of *Brassica campestris*. The *B. campestris* cytoplasm leads to the following cytoplasmic characters: normal green leaves (normal), well developed nectaries (N^+), atrazine resistance (Atr^r) and male fertility (F).

Experimental conditions used in protoplast isolation, protoplast fusion and plant regeneration have been detailed previously (Pelletier *et al.*, 1983). In a first experiment, protoplasts from "Brutor" and "C" varieties were fused: 5 cybrids (lines 23, 27, 58, 85, 118), combining *B. napus* chloroplasts and *R. sativus* cms trait, were screened among 176 regenerated plants. In a second experiment, protoplasts from "C" and "Tower" varieties were fused: one cybrid (line 77) combining chloroplasts fo triazine resistant *B. campestris* and *R. sativus* cms trait was detected among 199 regenerated plants (Table I). Each cybrid was crossed to a "Brutor" ("B" variety) to obtain the first and successive progenies. The restorer line R was derived from the material obtained by Heyn (1978) who selected it from a cross between a cms O *B. napus* as female and a *B. napus* X *R. sativus* amphidiploic as male. The R line is characterized by 38 chromosomes and white fertile flowers. Cms O lines of *B. napus* segregated fertile and sterile plants when pollinated by the restorer line. The precise genetic constitution of the R line and the determinism of restoration was not fully elucidated.

MITOCHONDRIAL DNA POLYMORPHISM IN RAPESEED CYBRIDS

The mt-DNAs of the parental lines diverged in restriction site markers so that restriction fragments unique to each parent could be identified. Fig. 1 shows the *Sal* I restriction patterns for mt-DNAs from "B" and "C" parents showing 22 and 24 bands, respectively; 14 of these appear to be common to "B" and "C" mt-DNAs, 8 specific to "B" (fertile line) mt-DNA and 10 specific to "C" (cms line) as shown by comparison of "B" and "C" schematic diagrams. Fig. 1 also presents the *Sal* I patterns of mt-DNAs

isolated from the progeny (fourth generation) of the six regenerated
cybrids, obtained by crossing female cybrid x male "Brutor". Each of the

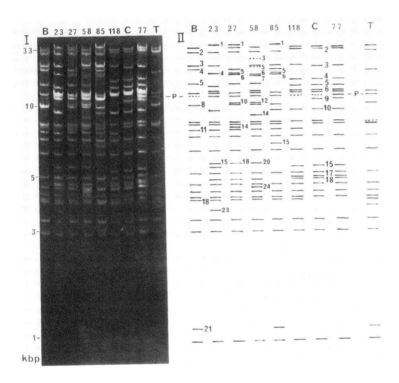

Fig. 1 – I. Electrophoresis on 0.7% agarose gel of *Sal* I
digests of mt-DNA isolated from parents of the two
fusion experiments and from their cybrids (B) "Brutor"
variety a fertile line of rapeseed, (C) *cms* O line of
rapeseed with a radish cytoplasm; (23), (27), (58), (85),
(118) cybrids from "Brutor" and "C"; (T) "Tower"
variety a fertile line of rapeseed with Atr^r *B. campestris*
cytoplasm; (77) cybrid from "C" and "Tower". II. Schematic
representation of the different *Sal* I restriction patterns.
Specific fragments of "Brutor" and "C" lines and new
fragments in cybrids are numbered according to their
decreasing molecular weight. P, locates the 11.5 pkb
mitochondrial plasmid like DNA. Digestion with *Sal* I
enzyme converts the main mitochondrial genome into a series
of discrete bands without affecting the mitochondrial plasmid.

Table I Cytoplasmic genotypes and phenotypes of parent and cybrid plants

		Organelle DNA		Cytoplasmic traits			
		Chloroplast	Mitochondria	Atrazine	Chlorophyll	Nectary	cms/F
Parents	"Brutor"	*B. napus*	*B. napus*	Atr^s	normal	N^+	F
	"C"	*R, sativus*	*R. sativus*	Atr^s	deficient	N^-	cms O
	"Tower"	*B. campestris*	*B. campestris*	Atr^r	normal	N^+	F
Cybrids	23	*B. napus*	described	Atr^s	normal	N^+	cms
	27	*B. napus*	in this	Atr^s	normal	N^+	cms
	58	*B. napus*	paper	Atr^s	normal	N^+	cms
	85	*B. napus*		Atr^s	normal	N^+	cms
	118	*B. napus*		Atr^s	normal	N^+	cms
	77	*B. campestris*		Atr^r	normal	N^+	cms

Atr^s – atrazine-susceptible: Atr^r = atrazine-resistant; N^+ = normal nectaries; N^- = underdeveloped nectaries; F = male fertile; cms "O" = cytoplasmic male sterility discovered in radish by Ogura (1968) and transferred into *Brassica* by Bannerot et al., (1974). "Brutor", "C" and "Tower" are spring varieties.

five cybrids corresponding to the first protoplast fusion experiment contains a unique mt-DNA distinguishable from both parents mt-DNAs and from a mixture of the two. However, the cybrid patterns always contain characteristic fragments of both parents and in all but one (cybrid 118) new bands appear (Table 2). The only cybrid (77) of the second fusion experiment possess solely the "C" mt-DNA as evidenced from *Sal* I (Fig. 1) and *Bgl* I and *Kpn* I (not shown) patterns. This line is atrazine resistant and possess the *B. campestris* ct-DNA type (Table I). Restriction patterns of cybrids 27, 58 and 85 are characterized by several new fragments of varying stoichiometries precluding accurate determination of molecular weight. However, estimates obtained by adding up the molecular weights of the bands (the plasmid band excepted) show that the cybrid mt-DNAs more closely resemble the "C" mt-DNA than the "Brutor" mt-DNA (Table II). Table II also shows for each cybrid mt-DNA the distribution of the specific and common parental bands. A correlation exists between the ratio of specific *B. napus* and *R. sativus* bands and the degree of sterility as evidenced from the restoration experiments (see below).

Mt-DNA polymorphism of the cybrids may also be illustrated from mitochondrial plasmid variation analysis and from gene mapping experiments. Palmer, Shields, Cohen and Corton, (1983) described in some *Brassica* species an unusual mt-DNA plasmid, 11.5 kbp in size, important quantitative variations of which occur in both parental and somatic hybrid plants (Chetrit *et al.*, 1985, see also Fig. 1). Our previous result suggests that there is no causal relationship between mitochondrial plasmid and cms. Differences between the N ("Brutor") and cms ("C") lines of *B. napus* are evidenced in the autoradiographs of Fig. 2 by comparing the location of the mitochondrial genes for cytochrome oxidase subunit II (*COII*) and ribosomal RNA. In Fig. 2 A, the nick-translated maize *COII* probe (Fox and Leaver 1981) was hybridised with parental and cybrid *Sal* I mt-DNA restriction fragments. *COII* gene location appears identical in cybrids 23, 27, 58, 85 and in fertile Brassica parents on the one hand in cybrids 118, 77 and in the cms radish parent on the other hand. In Fig. 2b, the *B. oleracea* mt-DNA *Sal* I fragment, S6 carrying the 18S and 5S mt-rDNA genes (Chetrit *et al.*, 1984) was nick-translated and hybridised with parental and cybrid *Sal* I mt-DNA patterns. The 18S and 5S gene location appears similar in cybrids 27, 58, 85 and fertile *Brassica* parents (S6 of *B. oleracea* being

Table II *Sal I* digests of mt-DNA

Plants	Common	B. napus	B. sativus –	New [a]	Total	MW [b] (kbp)
			bands			
B	14	8	–	–	22	222
23	13	4	5	4	26	250
27	14	6	1	6	27	281
58	14	6	2	8	30	320
85	13	7	1	4	25	260
118	13	3	8	–	24	232
C and 77	14	–	10	–	24	238

Cytoplasmic genotypes and phenotypes of parent and cybrid plants are defined in Table I.

(a) *Sal* I fragments with a new molecular weight as indicated in Fig. 1

(b) Approximate molecular weights obtained as described in the text.

homologous to S8 of *B. napus*) on the one hand and in cybrids 23, 118, 77 and cms parent (S6 of *B. oleracea* being homologous to S5 of *R. sativus*) on the other. S3, the *B. oleracea Sal* I fragment carrying the mt-26S rDNA leads to similar hybridisation patterns with cybrid 23, 118, 77 and cms parental mt-DNAs (not shown). Each of the cybrid 27, 58, 85 and fertile rapeseed lines may be distinguished with S3. This last result suggests that the 26S rDNA gene is involved in recombinational events induced by protoplast fusion. Naturally occurring recombinational events implicating rDNA genes have been evidenced recently in *B. oleracea* (Chetrit *et al.*, 1984) and in *T. aestivum* (Falconet *et al.*, 1984; Quetier *et al.*, 1985).

EVIDENCE FOR INTERGENOMIC MITOCHONDRIAL RECOMBINATION DURING PROTOPLAST FUSION

The novel fragments detected in cybrids mt-DNA patterns were interpreted as physical evidence of molecular recombination. In an attempt to verify this hypothesis, we have (1) prepared the largest novel fragment from

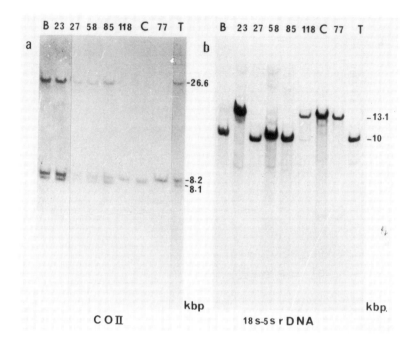

Fig. 2 - Mt-DNA *Sal* I patterns from parents "Brutor" and "C" and their
five somatic hybrids and from parents "C" and "Tower" and
their only somatic hybrid were blotted and hybridised to
nick-translated maize *COII* probe (Fox and Leaver 1981) (a)
and cauliflower *Sal* I fragment, 56, carrying the 18S and 5S
mitochondrial rDNA genes (b).

cybrid 27 mt-DNA and (2) analysed this band to detect sequences specific
of each parent. The novel band 1 (as with the other five novel bands) was
amplified from a clone library prepared by inserting *Sal* I incomplete
restriction digests of cybrid 27 mt-DNA into the pHC 79 cosmid vector.
This band was nick-translated and hybridised to blots of parental and
cybrid mt-DNA *Sal* I patterns. The autoradiographs of Fig. 3 show
homologies between the new band 1 and band 1 of the "Brutor" line ("B")
and bands 1 and 6 of the cms line ("C"). In a second step, these three
parental fragments were isolated by preparative electrophoresis of "B"
and "C" mt-DNA *Sal* I digests. Cybrid 27 new band 1 and each parental band
were digested with different restriction enzymes and the digests separated

by electrophoresis. The corresponding blots were hybridised to the nick-translated new band 1 probe. Autoradiographs with *Pst* I and *EcoRI* (not shown) indicate that fragment 1 from "Brutor" and fragment 6 from "C" mt-DNAs constitute the cybrid 27 band 1. Characteristics of band 1 in cybrid 23 and cybrid 85 (molecular weight identical to band 1 in cybrid 27 and hybridization with cybrid 27 band 1 probe as shown in Fig. 3) suggest that these bands have the same composition as cybrid 27 band 1. While we have strong evidence that cybrid mt-DNA restriction patterns contain both fragments specific of each parent and novel fragments with a hybrid composition, the nature of the recombination which has occurred has still to be investigated.

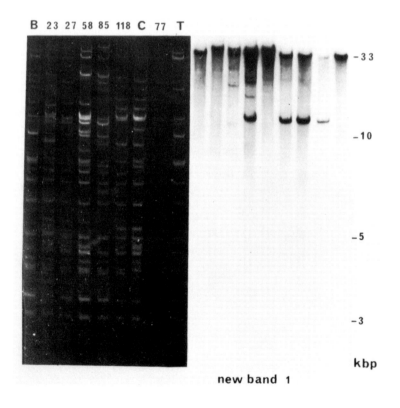

new band 1

Fig. 3 - Hybridisation of the mt-DNA *Sal* I patterns from parents of fusion and somatic hybrids to the cybrid 27 [32]p labelled novel band 1. Left are the gel patterns and right, the corresponding autoradiographs.

The mitochondrial genomes of the *Brassica* species being approximately 220 kb in size are among the smallest in higher plants (Lebacq and Vedel, 1981). Their simplicity has allowed the physical mapping of *B. campestris* (Palmer and Shields, 1984) and *B. oleracea* (Chetrit *et al.*, 1984) mt-DNAs. Each genome can be organised as a single circular molecule containing the entire sequence complexity. In *B. campestris*, this master circle can segregate into two smaller circles by reciprocal recombination within a major repeat element. However, in *B. oleracea*, our results suggest a greater number of recombinational events than that predicted in *B. campestris*. If repeat elements are in part responsible for naturally occurring intragenomic recombination in *Brassica* species, it is likely that these elements play an important role in protoplast fusion induced intergenomic recombination. Fig. 4 shows hybridisation between parent and cybrid mt-DNA *Sal* I patterns and a *B. oleracea* mt-DNA *Bgl* I fragment (B18, 2.5 kbp in size bearing a repeat element, Chetrit *et al.*, 1984) used as a probe. Band 1 from both "Brutor" and "C" lines is unlabelled as opposed to fragment 2. In the same manner, fragment 6 from the "C" line, appears unlabelled as opposed to fragment 7. The parental bands of the cybrid 27 fragment 1 do not contain the major repeat sequence characteristic of *Brassica* species. If repeat sequences mediate reciprocal recombination leading to cybrid 27 fragment 1, they should be located on fragments adjacent to "Brutor" fragment 1 and "C" fragment 6. Physical mapping experiments with *B. napus* and *R. sativus* mt-DNAs are in progress to elucidate this point.

Hybridisation patterns in Fig. 4 strongly resemble those obtained with the maize *COII* probe in Fig. 2a, corroborating the recent finding by Palmer (1984) that the major repeat in *B. campestris* mt-DNA contains the *COII* gene.

NATURAL RECOMBINATION AND PROTOPLAST FUSION-INDUCED RECOMBINATION OCCUR
AT HOMOLOGOUS SITES IN THE MITOCHONDRIAL GENOMES
According to the model of Palmer and Shields, the 2 kbp repeats are supposed to play a role in natural recombination, a process by which the master chromosome is dispersed into sub-molecules. The relative positions of the repeats coupled with relative recombination frequency will dictate the genome organisation. A high percentage of cauliflower mitochondrial molecules was found by electron microscopy to be about 40-45 kb in size, a distribution that can also be predicted from the physical map (Chetrit

et al., 1984). Highly homologous sequences of the cauliflower genome are located on five regions involving the *Sal* I fragments numbers 1, 2b, 3, 7, 8a and 15. Fragments 2b, 7 and 8a are homologous respectively to *B. napus Sal* I fragments numbers 2, 9 and 10 which appear labelled in Fig. 4.

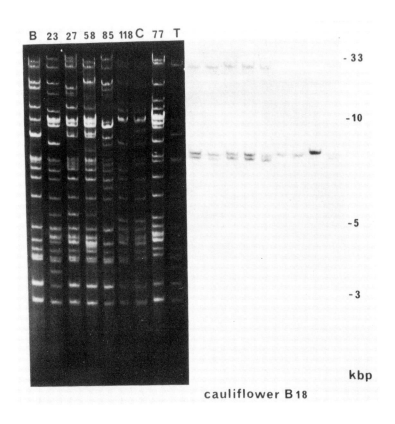

cauliflower B18

Fig. 4 - Hybridisation of the mt-DNA *Sal* I patterns from parents of fusion and somatic hybrids with the cauliflower ^{32}p labelled fragment B18 bearing a 2 kbp major repeat. Shown on the left, are the gel patterns and right, the corresponding autoradiographs.

The construction of *B. oleracea* mt-DNA physical maps was greatly simplified by the high mt-DNA content in the cauliflower inflorescences (Vedel and Mathieu, 1982). To overcome the difficulty of low mt-DNA content in the leaves of fertile or cms *B. napus* and because *B. oleracea* is

one of the rapeseed natural parents, the first approach tried was to
determine the sequential homology between *B. oleracea* and fertile and cms
B. napus Sal I fragments. Each *B. oleracea Sal I* mt-DNA band was isolated,
nick-translated and hybridised to parents and cybrids *Sal I* patterns.
Another complementary approach consisted in the hybridisation between each
cybrid 27 cloned new band (as indicated in Fig. 3 with band 1) and parents
and cybrids patterns.

If these experiments do not allow the construction of complete *Sal I*
map of *B. napus* mt-DNAs, they help to determine how mt-DNA recombination
has occurred in cybrids. In lines 23, 27, 58, 85 and 118, *Sal I* fragments

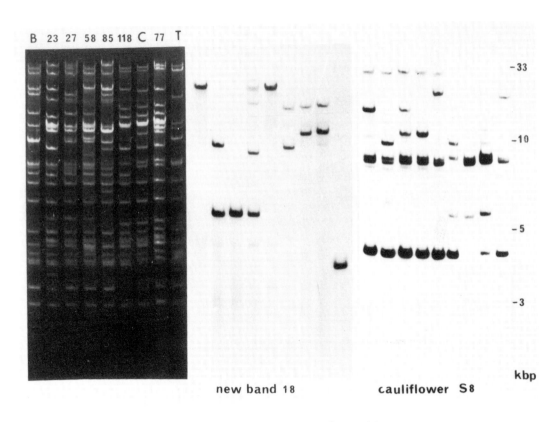

new band 18   cauliflower S8   kbp

Fig. 5 - Mt-DNA *Sal* I patterns from parents and somatic
hybrids (left) were blotted and hybridised to nick-
translated *Sal* I new band 18 from cybrid 27 mt-DNA
(centre) and *Sal* I band 8 from cauliflower mt-DNA
(right).

203

with a new molecular weight or a new hybridization pattern were considered
as specific of cybrids and resulting from intergenomic mitochondrial
recombination. *Sal I* fragments with a new molecular weight are indicated
in Fig. 1. Examples of *Sal I* fragments with a new hybridisation pattern
are given in Fig. 5 with cybrid 27 cloned new band 18 and cauliflower
band 8 probes. The key features are (1) "Brutor" and cybrid 85 present
identical patterns with the labelled fragment 3, (2) identical patterns
are observed in "C" and cybrid 77 lanes, bands 4 and 9 of which are
labelled and (3) cybrids 23, 27, 58 and 118 are characterized by unique
autoradiographs. New band 18 hybridizes with itself only in cybrid 27
pattern and with two bands in cybrids 23, 58 and 118. In cybrid 58 pattern,
bands 14 (a new one) and 20 (the same size as the probe) appear labelled.
Labelled fragments 9 in both cybrids 23 and 118 autoradiographs have
identical sizes. It is significant that fragment 10 in "C" (or 77) pattern
is not labelled although it is of the same size as fragment 9 in both
cybrids 23 and 118. On the basis of their unique labelling, fragments 9
are also considered as novel fragments. Cauliflower band 8 probe also
leads to specific autoradiographs (Fig. 5 right). However, these patterns
appear complex because band 8, isolated from a mt-DNA *Sal I* digest by
preparative electrophoresis, consisted of two distinct fragments, one of
which bears a 2 kbp major repeat.

Thirty-eight novel fragments in all were detected. Among them, is
hybridise with *B. oleracea* fragment 3, 10 with fragment 2b; 6 with
fragment 8a; 6 with fragment 15, 6 with fragment 1; 2 with fragment 7.
These fragments fit well with fragments corresponding to or adjacent to
natural recombination sites in cauliflower mt-DNA (Fig. 6).

The main conclusion is that mt-DNA recombination appears site specific.

More precisely, hybridisation experiments have shown that "Brutor" fragment 1 and C fragment 6 (contained in cybrid 27 novel fragment 1) are
respectively colinear with cauliflower fragments 1 and 15. Flanking
sequences of these cauliflower fragments bear a repeat (Fig. 6) suggesting
that the 2 parental fragments in cybrid 27 new band 1 are associated by
recombination at identical repeats in radish and rapeseed mt-DNAs. Furthermore, these results suggest that mitochondrial rDNA genes are involved in
protoplast fusion-induced molecular events since cauliflower fragment 3
bears the 26S *rRNA* gene and cauliflower fragment 8a (bearing a 2 kbp repeat)

is adjacent to fragment 6 carrying the 18S and 5S *rRNA* genes.

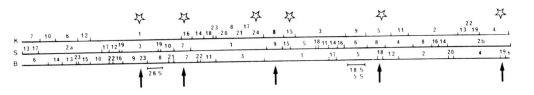

Fig. 6 - Restriction maps of cauliflower mt-DNA restricted
with *Kpn* I (K), *Sal* I (S) and *Bgl* I (B) enzymes,
indicating fragments involved in natural recombination
(arrows) and fragments hybridising with novel bands
arising from mt-DNA recombination in cybrids (stars).
The circular map containing the entire sequence
complexity of the genome was linearized by a cut at the
Sal I site internal to B6 and K7 fragments. Genes coding
for 26S, 18S and 5S mt-rDNA were located on this map
(Chetrit *et al.*, 1984).

FERTILITY RESTORATION OF CYBRIDS

Mitochondrial DNA recombination was also demonstrated using a classical
genetic approach. In general, a cms character is qualitatively defined by
a specific genetic (nuclear) system of fertility restoration. The original
cms "O" radish cytoplasm used in this study is restored by genes existing in
the radish nucleus. To analyse the fertility restoration of these new cms
cybrids, crosses were made with a *B. napus* line possessing restorer genes
from radish (R) in a heterozygous state. Table III contains the results
of this study. The original cms O radish cytoplasm as well as cybrid
cytoplasms (27, 58, 85, 118, 77) were restored by the R line, confirming
that cybrids actually retained the male sterility character present in the
cms parent of fusion. These results also ruled out two other possibilities.
i.e. *in vitro* induced variation led to cms and that a new cms character
was created by recombination between parental mt-DNAs.

Table III   Differential segregation into fertile (F), 1/2 fertile and
             sterile (S) plants among the progeny of sterile plants bearing
             cybrid cytoplasms (27, 58, 85, 118, 77) or an Ogura radish
             cytoplasm ("O") crossed by either a maintainer line ("M") or
             a heterozygous restorer line ("R") and among the progeny
             of self-pollinated restored plants.

Crossing	Flower phenotype		
	F	1/2 F	S
27 x "M"	0	0	10^4
27 X "R"	44	0	56
(27 X "R") selfed	80	0	19
58 X "M"	0	0	10^4
58 X "R"	43	0	54
(58 X "R") selfed	81	0	20
85 X "M"	0	0	10^4
85 X "R"	33	0	27
(85 X "R") selfed	65	0	21
118 x "M"	0	0	10^4
118 X "R"	17	5	27
(118 X "R") selfed	65	26	55
77 X "M"	0	0	10^4
77 X "R"	12	6	39
(77 X "R") selfed	25	14	37
"O" X "M"	0	0	10^4
"O" X "R"	22	11	65
("O" X "R") selfed	32	18	46

In the case of O, 77 and 118 cytoplasms, the progeny segregate into
three different phenotypes : fertile, half fertile and sterile. So, at
least two restorer genes are implicated. In the case of cybrids 27, 58,

85, the progeny segregates only into two different phenotypes, and these ratios are in agreement with the need for only one restorer gene to restore fertility.

In order to interpret these results, taking into account mt-DNA recombination as shown by molecular analysis, it must be assumed that "O" mt-DNA bears more than one cms factor and that 27, 58 and 85 recombinants possess fewer cms factors. The simplest hypothesis is to consider that "O" mitochondria leads to cms for two independent reasons: the "Ogura" male sterility already expressed in radish (Ogura, 1968) and the alloplasmic male sterility expressed when radish mitochondria are in the presence of a *Brassica* nucleus (Mac Callum, 1981). Mt-DNA recombination makes it possible to separate genetically these two factors and to give a simpler system of restoration in some cases (cybrids 27, 58, 85).

Fertility restoration results are in agreement with *COII* and *18S - 5S rDNA* gene mapping (Fig. 2) leading to the distinction of two classes of cybrids : 27, 58, 85 on the one hand and 118, 77 on the other. Fertility restoration experiments are currently in progress with the hybrid 23. At present, it is not known whether or not these genes are involved in cytoplasmic male sterility.

In fact, this male fertility restoration system has raised an unexpected problem rendering it, up to now, of no use for hybrid seed production. Among segregating progeny of self-pollinated restored plants, a dramatic decrease in female fertility is observed. If the restorer genes are actually the genes responsible for this female sterility, this observation would be of great interest in the understanding of the physiology of sterility.

PROSPECTS AND CONCLUSION

The comparison between physical maps of parent and cybrid mt-DNAs could make it possible to localize the cms factors. To reach this goal, the following materials need to be obtained:

(1) a sterile plant and a fertile plant differing only by these particular genes. The current strategy consists of carrying out back-fusion experiments in which cms cybrids are fused with fertile plants in order to get an almost

normal (*B. napus*) mitochondrial genome with the cms genes only from radish and in the reciprocal situation by fusing fertile cybrids with plants bearing "O" cytoplasm to get a radish mitochondrial genome with the fertility genes of only *B. napus*, and (2) physical maps of *B. napus* and *R. sativus* mt-DNAs using the methods previously used to construct the califlower mt-DNA maps. Hybridization experiments between nick-translated cauliflower mt-DNA fragments and blots of *B. napus* and *R. sativus* mt-DNA restriction patterns lead to incomplete physical mapping.

In conclusion, it is possible in *Brassica napus* to create new cytoplasmic associations which are of great interest for plant breeding. The material we have obtained is now used by breeders of rape and cabbage.

This material also offers the opportunity to understand better the mechanism of mt-DNA recombination which appears to be site specific, and in the future, to know which genes are implicated in the cms phenomenon.

REFERENCES

Bannerot, H., Boulidard, L., Cauderon, Y. and Tempe, J. (1974). Transfer of cytoplasmic male sterility from *Raphanus sativus* to *Brassica oleracea*. *Proc. Eucarpia Meeting Cruciferae*, pp. 52 - 54.

Belliard, G., Pelletier, G., Vedel, F. and Quetier, F. (1978). Morphological characteristics and chloroplast DNA distribution in different cytoplasmic parasexual hybrids of *Nicotiana tabacum*. *Mol. Gen. Genet. 165* : 231 - 237.

Belliard, G., Vedel, F. and Pelletier, G. (1979). Mitochondrial recombination in cytoplasmic hybrids of *Nicotiana tabacum* by protoplast fusion. *Nature 281* : 401 - 403.

Boeshore, M.L., Lifshizt, I., Hanson, M.R. and Izhar, S. (1983). Novel composition of mitochondrial genomes in *Petunia* somatic hybrids derived from cytoplasmic male sterile and fertile plants. *Mol. Gen. Genet. 190* : 459 - 467.

Chetrit, P., Mathieu, C., Muller, J.P. and Vedel, F. (1984). Physical and gene mapping of cauliflower *(Brassica oleracea)* mitochondrial DNA. *Current Genet. 8* : 413 - 421.

Chetrit, P., Mathieu, C., Vedel, F., Pelletier, G. and Primard, C. (1985). Mitochondrial DNA polymorphism induced by protoplast fusion in Cruciferae. *Theor. Appl. Genet. 69* : 361 - 366.

Duvick, D.N. (1965). Cytoplasmic pollen sterility in corn : *Adv. Genet. 13 :* 1 - 56.

Falconet, D., Lejeune, B., Quetier, F. and Gray, M.W. (1984). Evidence for homologous recombination between repeated sequences containing 18S and 5S ribosomal RNA genes in wheat mitochondrial DNA. *EMBO J., 3 :* 297 - 302.

Fluhr, R., Aviv, D., Edelman, M. and Galun, E. (1983). Cybrids containing mixed and sorted out chloroplasts following interspecific somatic fusions in *Nicotiana. Theor. Appl. Genet. 65 :* 289 - 294.

Fox, T.D., and Leaver, C.J. (1981). The *Zea mays* mitochondrial gene coding cytochrome oxidase subunit II has an intervening sequence, and does not contain TGA codons. *Cell, 26 :* 315 - 323.

Galun, E., Arzee-Gonen, P., Fluhr, R., Edelman, M. and Aviv, D. (1982). Cytoplasmic hybridization in *Nicotiana :* mitochondrial DNA analysis in progenies resulting from fusion between protoplasts having different organelle constitution. *Mol. Gen. Genet. 186 :* 50 - 56.

Heyn, F.W. (1978). Introgression of restorer genes from *Raphanus sativus* into cytoplasmic male sterile *Brassica napus* and the genetics of fertility restoration. In: *Proc. 5th Int. Rapeseed Conf. Malmoe,* pp. 82 - 83.

Lebacq, P. and Vedel, F. (1981). *Sal* I restriction enzyme analysis of chloroplast and mitochondrial DNAs in the genus *Brassica. Plant Sci. Lett. 23 :* 1 - 9.

Lonsdale, D.M., Hodge, T.P. and Fauron, C.M.R. (1984). The physical map and organization of the mitochondrial genome from the fertile cytoplasm of maize. *Nucleic Acids Res. 12 :* 9249 - 9261.

Mc. Collum, G.D. (1981). Induction of an alloplasmic male sterile *Brassica oleracea* by substitution cytoplasm from "early scarlet globe" radish (*Raphanus sativus*). Euphytica *30.* : 855 - 859.

Nagy, F., Torok, I. and Maliga, P. (1981). Extensive rearrangements in the mitochondrial DNA in somatic hybrids of *Nicotiana tabacum* and *Nicotiana knightiana. Mol. Gen. Genet. 183 :* 437 - 439.

Ogura, H. (1968). Studies on the new male sterility in radish with special reference to the utilization of this sterility towards the practical raising of hybrid seeds. *Mem. Fac. Agric. Kagoshima University, 6 :* 39 - 78.

Mitochondrial genomes

Palmer, J.D., Shields,C.R., Cohen, D.B. and Orton, T.J. (1983). An unusual
    mitochondrial DNA plasmid in the genus *Brassica*. *Nature 301* : 725 - 728.

Palmer, J.D., (1984). Organization and evolution of mitochondrial DNA in
    Brassica. *EMBO workshop on Plant mitochondrial DNA*, Melrose.

Palmer, J.D. and Shields, C.R. (1984). Tripartite structure of the
    Brassica campestris mitochondrial genome. *Nature, 307* : 437 - 440.

Pelletier, G., Primard, C., Vedel, F., Chetrit, P., Remy, R., Rousselle, P.,
    and Renard, M. (1983). Intergeneric cytoplasmic hybridization in
    Cruciferae by protoplast fusion. *Mol. Gen. Genet. 191* : 244 - 250.

Quetier, F., Lejeune, B., Delorme, S. and Falconet, D. (1985). Molecular
    organization and expression of the mitochondrial genome of higher plants
    in *Encycl. Plant Physiol. New Series Vol. 18, Higher plant Cell
    Respiration* (eds. Douce, R. and Day, D.A., Verlag, Berlin, Springer -
    25 - 36.

Vedel, F. and Mathieu, C. (1982). Isolation of purified mitochondrial DNA
    from *Brassicae*. *Anal. Biochem. 127* : 1 - 8.

Significance of the chondriome to plant breeding

10   Cybrids as tools in higher plant genetics

G. Pelletier

ABSTRACT

Somatic hybridization in higher plants allows the mixing of organelles in the same cell. Selection or screening methods have been exploited in order to detect cytoplasmic genome exchanges and genome recombinations between parental organelles. Mitochondrial DNA recombination appears to be a general phenomenon observed in the *Solanaceae*, brassicas, and *Umbelliferae*. Somatic mixing would permit a better understanding of the expression and organisation of this particularly large genome of higher plants. Chloroplast DNA recombination has been also recently described, rendering the cell fusion technic a complete tool for plant cytoplasmic genetics in which it would play a major role in the future.

INTRODUCTION

Polyploidy occurs widely in the plant kingdom, and in cultivated plants there exist autoployploids, e.g. potato and alfalfa, while others are allopolyploid e.g. wheat, oilseed rape and tobacco. Allopolyploidy is the result from natural crosses between related species and from chromosome doubling. For example, wheat was spontaneously produced by crosses between *Triticum* and *Aegilops* species; rapeseed came from the summation of cabbage and turnip genomes. If such additions of nuclear genomes are sexually possible, sometimes with the aid of more sophisticated procedures like embryo culture, the addition or recombination of cytoplasmic genetic information is also. Naturally, this is impossible because the fertilization process prevents the participation of the male cytoplasm, (see Russell, this volume).

   Induced fusion of isolated protoplasts followed by regeneration of entire plants allows the mixing of cytoplasms and the screening or the selection of recombined forms of cytoplasmic genetic characteristics. These forms are called "cybrids" in the cases of cytoplasmic hybrids.

The chondriome and plant breeding

Studies on cybrids began about ten years ago (Gleba, Butenko and Sytnik, 1975) with model species belonging to the *Solanaceae*. During this period, several sophisticated methods were tried to improve the cybrid ratio obtainable per experiment. But the current interest in cybrids resides now in new genetic results concerning mitochondrial and chloroplast genomes which have never before been obtained in higher plants.

METHODS FOR OBTAINING CYBRIDS

Protoplast fusion implies cytoplasmic mixing but does not necessarily lead to nuclear fusion. Therefore among fusion products at the level of regenerated plants, several possess one or the other unmodified parental nucleus and a hybrid cytoplasm which can be screened if suitable genetic markers are used. This is the simplest procedure to obtain cybrids.

Nevertheless, several workers have tried to improve the process of cybrid recovery by using more sophisticated methods. Zelcer, Aviv and Galun (1978) used X-rays to irradiate protoplasts, and so destroy their nucleus. Such protoplasts were unable to divide further and could participate only by their cytoplasm in fusion products. Sidorov, Menczel, Nagy and Maliga (1981) introduced an additional refinement by treating the other partner with iodoacetate, a compound acting as a metabolic inhibitor of the cytoplasm. Maliga, Lorz, Lazar and Nagy (1982) used cytoplasts (mechanically enucleated protoplasts) for preferentially transfering one type of cytoplasm.

In all probability, in fact, the previously proposed procedures are either of variable efficiency and somewhat complicated or genetically confusing when inactivation procedures are employed. The most suitable methods are based upon genetic screening of regenerated plants (Belliard, Pelletier and Ferault, 1977; Pelletier, Primard, Vedel, Chetrit, Remy, Rousselle and Renard, 1983) or isolation of fused cells in culture in those cases where morphological differences between parental protoplasts allow it (Gleba, Kolesnik, Meshkene, Cherep, and Parokonny, 1984) or selection based on antibiotic resistances (Medgyezy, Menczel and Maliga, 1980).

THE FATE OF CHLOROPLAST GENOMES AFTER PROTOPLAST FUSION

Genetic markers are available which permit the screening or the selection *in vitro* or at the plant level of one type of chloroplast genome. The most

214

commonly occurring markers are chlorophyll-deficient mutants. In fusion with protoplasts, possessing normal plastids, the latter can easily be recovered by selecting green calli or shoots (Gleba, Pisen, Komarnitsky and Sytnik, 1978).

Mutants displaying resistances can be recovered after positive selection. Examples of which are resistance to tentoxin (Zelcer *et al.*, 1978), streptomycin (Medgyezy *et al.*, 1980), lincomycin (Cseplo and Maliga, 1982) and triazine (Binding, Jain, Fuiger, Wordhorst, Nehts and Gressel, 1982; Pelletier *et al.*, 1983).

The identification of the chloroplast genome in order to confirm its parental origin is possible through DNA restriction pattern analysis (Belliard, Pelletier, Vedel and Quetier, 1978). This technique is generally applicable because differences exist between chloroplast DNA of different species. It also offers a relatively high resolution power (a one percent mixture can be detected) and allows detection of DNA rearrangements of the whole genome.

With attention to the fate of the chloroplast genomes after fusion, the first results (King, Gray, Wildman and Carlson, 1975; Chen, Wildman and Smith, 1977) showed, in the case of *N. glauca x N. longsdorfii* somatic hybrids, that a mixture of both chloroplast genomes seldom occurred at the plant level. Several further reports have confirmed these findings in somatic hybrids (with hybrid nuclei) as well as in cybrids (with one or the other nucleus): (Belliard *et al.*, 1978; Melchers, Sacristan and Holder, 1978; Aviv, Fluhr, Edelmann and Galun, 1980; Iwai, Nagao, Nakata, Kawastrima, Matsuyama , 1980; Poulsen, Porath, Sacristan and Melchera, 1980; Douglas, Wetter, Weller and Setterfield, 1981; Glimelius and Bonnet, 1981; Menczel, Nagy and Maliga, 1981; Scowcroft and Larkin, 1981; Schiller, Herrmann and Melchers, 1982). Analysis of the sorting-out of plastid types at the callus stage (Akada, Hirai and Uchimiya, 1983) confirms the hypothesis that it is achieved in the first few post-fusion mitoses. Furthermore, in some experimental systems it seems that chloroplast sorting-out could be unidirectional to one parental type (Flick and Evans, 1982; Kumar, Cocking, Bovenberg and Kool, 1982). A heteroplasmic state can sometimes be observed and proved by studying the progeny of a chimeral somatic hybrid if the progeny also contains chimeral plants. It can be inferred that cells with a mixed population of plastid genomes exist in the parent hybrid

The chondriome and plant breeding

(Gleba *et al.*, 1978; Fluhr, Aviv, Edelman and Galun, 1983; Fluhr, Aviv, Galun and Edelman, 1984; Gleba *et al.*, 1984). It is probably necessary to maintain the heteroplasmic state through several cell divisions in order to detect chloroplast DNA recombination after fusion. Until recently no convincing proof of recombination existed. However, by the use of biochemical mutants making it possible to detect very rare events, it was now possible to select the first definitive chloroplastic recombinant in higher plants (Medgyezy, Fejes and Maliga, 1985a).

THE FATE OF MITOCHONDRIAL GENOMES AFTER PROTOPLAST FUSIONS

The first system which made it possible to understand the fate of mitochondrial genomes after fusion was described in tobacco (Belliard *et al.*, 1977; 1978; 1979; Pelletier, Vedel and Belliard, 1985). The most pronounced case of male sterility known in tobacco, with a profound feminisation of the flower, is observed in an alloplasmic line of *N. tabacum* with *N. debneyi* cytoplasm. It is known that these cytoplasmic male sterile plants can be more or less restored by additional *N. debneyi* chromosomes, leading to a range of flower morphologies depending on the *N. debneyi* chromosome (Sand and Christoff, 1973). Protoplast fusion between this male sterile line and a normal *N. tabacum* line should give the opportunity of selecting phenotypically those plants of which the cytoplasm is a hybrid between *N. debneyi* and *N. tabacum*.

This experimental system made it possible to create several novel flower morphologies, intermediate between the parental types. Each flower type was strictly maternally inherited and stable through sexual generations. The study of ct-DNA confirms that these plants possessed either one or the other unmodified parental DNA, without any relationship to the flower morphology. On the contrary, the study of mt-DNA by restriction pattern analysis showed that each cybrid possessed its own pattern, combining bands specific to each parent with new bands. The conclusion of this study was that plants regenerated from protoplast fusions contain new types of mt-DNAs. created by recombination between parental mitochondrial DNA, and thus responsible for the new flower morphology. This event appeared to occur in the great majority of fusion products since it was also observed in the great majority of nuclear hybrids (56 out of 57).

The recombination of mt-DNA after protoplast fusion was further studied in other systems. (Nagy, Lazar, Menczel and Maliga, 1983) confirmed in *Nicotiana* that these rearrangements between parental mt-DNA are undetectable after homoplasmic fusion but commonly found after heteroplasmic fusions. In *Petunia*, Boeshore, Lifshitz, Hanson and Izhar (1983) and Hanson (1985) observed also novel compositions of mt-DNA in somatic hybrids. The studies of Nagy, Torok and Maliga (1981), and Galun, Arzee-Gonen, Fluhr, and Aviv, (1982) confirmed these results in other *Nicotiana* species. The results of Chetrit, Mathieu, Vedel, Pelletier and Primard (1985) and Matthews and Widholm (1985) widen the phenomenon to *Brassica* snd *Daucus* species respectively. Direct evidence that new fragments are actually the result of crossing over between sequences from a parent and sequences from the other parent were obtained in *Petunia* (Rothenberg, Boeshore, Hanson and Izhar, 1985) and *Brassica* (Vedel *et al.*, this volume).

Mitochondrial recombination after protoplast fusion is in accordance with physical maps of higher plant mt-DNAs which need such events to be explained (Palmer and Shields, 1984; Chetrit, Mathieu, Muller and Vedel, 1984; Lonsdale, Hodge and Fauron, 1984; Falconet, Delorme, Lejeune, Sevignac, Delcher, Bazeteux and Quetier, 1985). Homology between mitochondrial sites involved in naturally occurring recombination and mitochondrial sites involved in protoplast fusion induced recombination is shown by the results presented by Vedel *et al.*, (this volume) in cybrids obtained from protoplast fusion experiments in *Brassica*. From these results it seems that protoplast fusion induced mt-DNA recombination is mainly site specific just as is natural recombination.

There is good evidence that mitochondrial networks may be a feature of cells. They have been observed in apical cells (Walbot, Thompson, Keith and Coe, 1980), root tips (Gunning and Steer, 1975) and tissue culture cells (Rohr, 1978; 1979). Fusion of parental mitochondria reticula is probably a prerequisite to the mixing of genomes and ultimately recombinations between them. Nevertheless this step is still open to speculation because of the lack of ultrastructural studies during and after the fusion process detailing morphological rearrangements of intracellular membrane systems.

The chondriome and plant breeding

RELATIONSHIP BETWEEN THE THREE PLANT CELL GENOMES IN PROTOPLASTS FUSION
PRODUCTS

Plant cells contain three genomes: nuclear, chloroplastic and mitochondrial.
There is some evidence that common sequences are present in different
organelles. Homology between nuclear and chloroplast genomes (Timmis
and Scott, 1983), between chloroplast and mitochondrial genomes (Stern and
Lonsdale, 1982) and mitochondrial and nuclear genomes (Kemble, Mans, Gabay-
Laughnan and Laughnan, 1983) have been reported.

Functional relationship between these compartments can be investigated
by protoplast fusion experiments. Menczel, Nagy, Lazar and Maliga,(1983)
and Medgyezy, Golling and Nagy, (1985b) observed a preferential cotransfer
of ct-DNA (selected by streptomycin resistance or by normal greening
respectively) and cytoplasmic male sterility, a mitochondrial trait. These
results might suggest a cooperation between chloroplasts and mitochondria.
Another explanation could be that mt-DNA recombination occurs with a high
frequency and the mt-DNA would be (in all cases) of male sterile type by a
phenomenon similar to suppressiveness in certain petite mutants of yeast
(Locker, Lewin and Rabinowitz, 1979).

The cms system is a good example of genome interaction at the plant
level. Alloplasmic situations (i.e. those in which the nucleus of a given
species is introduced into the cytoplasm of a related one, often leads to
male sterility (Edwardson, 1970). Interaction between nuclear and chloroplast genomes in these situations are also found, leading to less effective
photosynthesis (Grun, 1976; Bannerot, Boulidard and Chupeau, 1977).

Protoplast fusion makes it possible to combine independently the nucleus,
chloroplast and mitochondria of different species. Pelletier *et al.,* (1983)
succeeded in improving a chlorophyll deficient-cytoplasmic male sterile
radish cytoplasm by substituting the chloroplast of *Brassica napus*. In the
same experiments a cybrid combining genomes from three different species was
produced: rapeseed nucleus, radish mitochondria containing genes encoding
for cms and turnip chloroplast containing genes encoding for triazine
resistance.

CONCLUSION AND PERSPECTIVES

Rapid and complete cytoplasmic exchange is possible between two lines by protoplast fusion. This is a great advantage compared to conventional methods (sexual) where this exchange is achieved after several generations of repeated back crosses. The cybrid method can compete with conventional methods only if good protoplast culture and plant regeneration procedures are available. Unfortunately, those are not available for all plant species and even not for all genotypes of an amenable species.

Different cytoplasmic traits can be gathered in a unique and stable cytoplasm. This is a totally new possibility for plant genetics now already being applied in plant breeding in the cases of *Brassica* and *Nicotiana*.

Organelle DNA recombination is now established for both chloroplast and mitochondrial genomes. The opportunity to create new hybrid chloroplast as well as mitochondrial genomes is already largely opened. Mitochondrial recombination offers the opportunity to locate sequences on cybrid restriction maps, involved in male sterility yet whose biochemical basis still remain unknown. Further information on these recombination phenomena would be of a great help to understand the molecular organisation of chloroplast and mitochondrial genomes. In particular, why differences occur and why higher plants possess such large and variable mitochondrial genomes, sometimes a hundred times larger than animals, when they apparently code for a similar number of proteins.

REFERENCES

Akada, S., Hirai, A. and Uchimiya, H.,(1983). Studies on mode of separation of chloroplast genomes in parasexual hybrid calli. I. Fraction I protein composition in unseparated hybrid callus. *Plant. Sci. Lett., 31* : 223 - 230.

Aviv, D., Fluhr, F., Edelman, M. and Galun, E. (1980). Progeny analysis of the inter-specific somatic hybrids : *Nicotiana tabacum* (*cms*) + *Nicotiana sylvestris* with respect to nuclear and chloroplast markers. *Theor. Appl. Genet., 56* : 145 - 150.

Bannerot, H., Boulidard, L. and Chupeau, Y. (1977). Unexpected difficulties met with the radish cytoplasm. *Eucarpia Cruciferae. Newsl., 2* : 16.

Belliard, G., Pelletier, G. and Ferault, M. (1977). Fusion de protoplastes de Nicotiana tabacum a cytoplasmes differents : etude des hybrides cytoplasmiques neo-formes. *C.R. Acad. Sc., 284* serie D. 749 - 752.

The chondriome and plant breeding

Belliard, G., Pelletier, G., Vedel, F. and Quetier, F. (1978).
Morphological characteristics and chloroplast DNA distribution in
different cytoplasmic parasexual hybrids of *Nicotiana tabacum. Mol. Gen.
Genet. 165* : 231 - 237.

Belliard, G., Vedel, F. and Pelletier, G., (1979). Mitochondrial recombination in cytoplasmic hybrids of *Nicotiana tabacum* by protoplast fusion.
Nature, 281 : 401 - 403.

Binding, H., Jain, S.M., Fuiger, J., Mordhorst, G., Nehls, R. and Gressel,
J., (1982). Somatic hybridisation of an atrazine resistant biotype of
Solanum nigrum with *Solanum tuberosum* : 1. Clonal variations in
morphology and atrazine sensitivity. *Theor. Appl. Genet. 63* : 273 - 277.

Boeshore, M.L., Lifshitz, I., Hanson, M.R. and Izhar, S., (1983). Novel
composition of mitochondrial genomes in Petunia somatic hybrids derived
from cytoplasmic male sterile and fertile plants. *Mol. Gen. Genet. 190* :
459 - 467.

Chen, K., Wildman, S.G. and Smith, H.H.,(1977). Chloroplast DNA distribution
in parasexual hybrids as shown by polypeptide composition of fraction I
protein. *Proc. Natl. Acad. Sci., USA., 74* : 5109 - 5112.

Chetrit, P., Mathieu, P., Muller, J.P., and Vedel, F., (1984). Physical
and gene mapping of cauliflower *(Brassica oleracea)* mitochondrial DNA
Curr. Genet. 8 : 413 - 421.

Chetrit, P., Mathieu, C., Vedel, F., Pelletier, G. and Primard, C., (1985).
Mitochondrial DNA polymorphism induced by protoplast fusion in
Cruciferae. Theor. Appl. Genet. 69 : 361 - 366

Douglas, G.C., Wetter, L.R., Weller, W.A., and Setterfield, G., (1981).
Somatic hybridisation between *Nicotiana rustica* and *N. tabacum.* IV
Analysis of nuclear and chloroplast genome expression in somatic
hybrids. *Can. J. Bot., 59* : 1509 - 1513.

Edwardson, J.R.,(1970). Cytoplasmic male sterility. *Bot. Rev. 36* : 341 - 420.

Falconet, D., Delorme, S., Lejeune, B., Sevignac, M., Delcher, E., Bazetoux,
S., and Quetier, F., (1985). Wheat mitochondrial 26S ribosomal RNA
gene has no intron and is present in multiple copies arising by
recombination. *Curr. Genet. 9* : 169 - 174.

Flick, C.E., and Evans, D.A., (1982). Evaluation of cytoplasmic segregation
in somatic hybrids of *Nicotiana* : tentoxin sensitivity. *J. Hered., 73* :
264 - 266.

Fluhr, R., Aviv, D., Edelman, M. and Galun, E., (1983). Cybrids containing mixed and sorted-out chloroplasts following interspecific somatic fusions in *Nicotiana*. *Theor. Appl. Genet., 65* : 289 - 294.

Fluhr, S., Aviv, D., Galun, E., and Edelman, M., (1984). Generation of heteroplastidic *Nicotiana* cybrids by protoplast fusion: analysis for plastid recombinant types. *Theor. Appl. Genet. 67* : 491 - 497.

Galun, E., Arzee-Gonen, P., Fluhr, R., Edelman, M., and Aviv, D., (1982). Cytoplasmic hybridization in *Nicotiana* : Mitochondrial DNA analysis of progenies resulting from fusion between protoplasts having different organelle constitutions. *Mol. gen. Genet. 186* : 50 - 56.

Gleba, Y.Y., Butenko, R.G., and Sytnik, K.M. (1975). Protoplast fusion and parasexual hubridization in *Nicotiana tabacum L. Doki. Acad. Nauk. USSR 221* : 1196 - 1198.

Gleba, Y.Y., Piven, N.M., Komarnitsky, I.K., and Sytnik, K.M., (1978). Cytoplasmic hybrids (cybrids) *Nicotiana tabacum + N. debneyi* obtained by protoplast fusion. *Dokl. Acad. Nauk. USSR., 240* : 1223 - 1226.

Gleba, Y.Y., Kolesnik, N.N., Meshkene, I.V., Cherep, N.N. and Parokonny, A.S., (1984). Transmission genetics of the somatic hybridization process in *Nicotiana*. *Theor. Appl. Genet., 69* : 121 - 128.

Gleba, Y.Y., Komarnitsky, I.K., Kolesnik, N.N., Meshkene, I., and Martyn, G.I., (1985). Transmission genetics of the somatic hybridisation process in *Nicotiana*. *Mol. Gen. Genet. 198* : 476 - 481.

Glimelius, K., and Bonnet, H.T., (1981). Somatic hybridization in *Nicotiana*. Restoration of photo autotrophy to an albino mutant with defective plastids. *Planta, 153* : 497 - 503.

Grun, P. (1976). Cytoplasmic genetics and evolution. Columbia University Press, New York.

Gunning, B.E.S. and Steer, M.W. (1975). Ultrastructure and the Biology of Plants Cells. Edward Arnold Ltd., London.

Hanson, M.R., (1984). Stability, variation and recombination in plant mitochondrial genomes via tissue culture and somatic hybridisation. In Oxford Surveys of Plant Molecular and Cell Biology. (ed. B.J. Miflin. Vol. 1. 33 - 52. Oxford University Press.

Iwai, S., Nagao, T., Nakata, K., Kawastrima, N., and Matsuyama, S., (1980). Expression of nuclear and chloroplastic genes coding for fraction I protein in somatic hybrids of *Nicotiana tabacum + N. rustica. Plants. 147* : 414 - 417.

Kemble, R.J., Mans, R.J., Gabay-Laughnan, S., and Laughnan, J.R. (1983). Sequences homologous to episomal mitochondrial DNAs in the maize nuclear genome. *Nature. 304* : 744 - 747.

Kumar, A., Cocking, E.C., Bovenberg, W.A. and Kool, A.J., (1982). Restriction endonuclease analysis of chloroplast DNA in interspecies somatic hybrids of *Petunia. Theor. Appl. Genet., 62* : 377 - 383

Kung, S.D., Gray, J.C., Wildman,S.G. and Carlson, P.S. (1975). Polypeptide composition of fraction I protein from parasexual hybrid plants in the genus *Nicotiana. Science. 187* : 353 - 355.

Locker, J., Lewin, A. and Rabinowitz, M. (1979). The structure and organization of mitochondrial DNA from petite yeast. *Plasmid. 2* : 155 - 181.

Lonsdale, D.M., Hodge, T.P., and Fauron, C.M.R., (1984). The physical map and organization of the mitochondrial genome from the fertile cytoplasm of maize. *Nucl. Acid Res., 12* : (24) 9249 - 9261

Maliga, P., Lorz, H., Lazar, G. and Nagy, F. (1982). Cytoplast-protoplast fusion for interspecific chloroplast transfer in *Nicotiana. Mol. Gen. Genet. 185* : 211 - 215.

Matthews, B.J., and Widholm, J.M. (1985). Organelle DNA composition and isoenzyme expression in an interspecific somatic hybrid of Daucus. *Mol. Gen. Genet. 198* : 371 - 376.

Medgyezy, P., Menczel, L. and Maliga, P. (1980). The use of cytoplasmic streptomycin resistance: chloroplast transfer from *Nicotiana tabacum* into *Nicotiana sylvestris* and isolation of their somatic hybrids. *Mol. Gen. Genet., 179* : 693 - 698.

Medgyezy, P., Fejes, E. and Maliga, P. (1985a). Interspecific chloroplast recombination in a *Nicotiana* somatic hybrid. *Proc. Natl. Acad. Sci. USA.* (In press).

Medgyezy, P., Golling, R. and Nagy, F. (1985b). Light sensitive recipient for effective transfer of chloroplast and mitochondrial traits by protoplast fusion in *Nicotiana. Theor. Appl. Genet.* (In press).

Melchers, G., Sacristan, M.D. and Holder, A.A. (**1978**). Somatic hybrid plants of potato and tomato regenerated from fused protoplasts. *Carlsberg. Res. Commun., 43* : 203 - 218.

Menczel, L., Nagy, F., Kiss, Z.R. and Maliga, P. (1981). Streptomycin resistant and sensitive somatic hybrids of *Nicotiana tabacum + Nicotiana knightiana* : correlation of resistance to *N. tabacum* plastids. *Theor. Appl. Genet., 59* : 191 - 195.

Menczel, L., Nagy, F., Lazar, G. and Maliga, P., (1983). Transfer of cytoplasmic male sterility by selection for streptomycin resistance after protoplast fusion in *Nicotiana. Mol. Gen. Genet. 189* : 365 - 369.

Nagy, F., Torok, I. and Maliga, P., (1981). Extensive rearrangements in the mitochondrial DNA in somatic hybrids of *Nicotiana tabacum* and *N. knightiana. Mol. Gen. Genet. 183* : 437 - 439.

Nagy, F., Lazar, F., Menczel, L. and Maliga, P. (1983). A heteroplasmic state induced by protoplast fusion is a necessary condition for detecting rearrangements in *Nicotiana* mitochondrial DNA. *Theor. Appl. Genet. 66* : 203 - 207.

Palmer, J.D. and Shields, C.R. (1984). Tripartite structure of the *Brassica campestris* mitochondrial genome. *Nature, 307* : 437 - 440.

Pelletier, G., Primard, C., Vedel, F., Chetrit, P., Remy, R., Rousselle, P. and Renard, M. (1983). Intergeneric cytoplasmic hybridization in Cruciferae by protoplast fusion. *Mol. Gen. Genet. 191* : 244 - 250.

Pelletier, G., Vedel, F. and Belliard, G. (1985). Cybrids in genetics and breeding. *Hereditas. Suppl. Vol. 3* : 49 - 56.

Poulsen, C., Porath, D., Sacristan, M.D. and Melchers, G. (1980). Peptide mapping of the ribulose biphosphate carboxylase small subunit from the somatic hybrid of tomato and potato. *Carlsberg. Res. Commun., 45* : 249 - 267.

Rohr, R. (1978). Existence d'un reticulum mitochondrial dans les cellules d'une culture de tissu haploide d'une plante vasculaire. *Bio. Cellulaire, 33* : 89 - 92.

Rohr, R. (1979). Reconstruction tridimensionnelle du chondriome de Ginkgo en culture tissulaire; etude au moyen de coupes epaisses et de coupes fines seriees. *Can. J. Bot., 57* : 332 - 340.

Rothenberg, M., Boeshore, M.L., Hanson, M.R. and Izhar, S. (1985). Intergenomic recombination of mitochondrial genomes in a somatic hybrid plant. *Curr. Genet. 9* : 615 - 618.

Sand, S.A. and Christoff, G.I. (1973). Cytoplasmic chromosomal interactions and altered differentiation in tobacco. *J. Hered. 64* : 24 - 30.

Schiller, B., Herrmann, R.G. and Melchers, G., (1982). Restriction endonuclease analysis of plastid DNA from tomato, potato and some of their somatic hybrids. *Mol. Gen. Genet. 186* : 453 - 459.

Scowcroft, W.R. and Larquin, P.J. (1981). Chloroplast DNA assort randomly in intraspecific somatic hybrids of *Nicotiana debneyi*. *Theor. Appl. Genet. 60* : 179 - 184.

Sidorov, V.A., Menczel, L., Nagy, F. and Maliga, P. (1981). Chloroplast transfer in *Nicotiana* based on metabolic complementation between irradiated and iodoacetate treated protoplasts. *Planta 152* : 328 - 332.

Stern, D.B., and Lonsdale, D.M. (1982). Mitochondrial and chloroplast genomes of maize have a 12-kilobase DNA sequence in common. *Nature 299* : 698 - 702.

Timmis, J.N. and Scott, N.S. (1983). Sequence homology between spinach nuclear and chloroplast genomes. *Nature. 305* : 65 - 67.

Walbot, V., Thompson, D., Keith, G.M. and Coe, E.H. (1980). Nuclear genes controlling chloroplast developments. In Genome organisation and expression in Plants (ed. Leaver, C.J.). 381 - 399. Plenum. New York.

Zelcer, A., Aviv, D. and Galun, E. (1978). Interspecific transfer of cytoplasmic male sterility by fusion between protoplasts of normal *Nicotiana sylvestris* and X-ray irradiated protoplasts of male sterile *N. tabacum*. *Z. Pflanzenphysiol. 90* : 397 - 407.

11 Elimination and substitution of chloroplasts in *Brassica napus*

C. H. Bornman, L. O. Björn, R. Vaňková, J. F. Bornman and C. Jarl

ABSTRACT

In alloplasmic cybrid construction it may be desirable to eliminate defective traits. One such trait is the chlorophyll deficiency associated with the transfer of the male sterility-conferring *Raphanus sativus* cytoplasm to *Brassica* spp.

We have attempted first to eliminate the offending *Raphanus* chloroplast genome in male-sterile *Brassica napus* and then, by cybridisation, to introduce a *Brassica* chloroplast genome. Chloroplasts were eliminated by (1) induction of a pathologic accumulation of protochlorophyllide using 5-aminolevulinic acid followed by irradiation with intense red light over the range 600-800 nm, (2) inhibition of carotenoid synthesis using norflurazon (SAN 9789) during germination and seedling development under constant light in the 670-680 nm range, and (3) direct irradiation with an helium-neon laser emitting at 632.8 nm.

By fusing chloroplast-free cytoplasts isolated from either petioles or hypocotyls of pre-sensitized cytoplasmically male sterile *Brassica napus* with protoplasts (or mini-protoplasts or nucleoplasts) isolated from the mesophyll of normal (male-fertile) *Brassica napus*, alloplasmic cybrid plants have been produced that are male sterile and, in respect of nucleus-chloroplast genomic interaction, also non-defective.

INTRODUCTION

Transfer of the gene that controls cytoplasmic male sterility cms from radish (*Raphanus sativus* L.) into the genus *Brassica*, including winter oil-seed rape or rapeseed (*Brassica napus* L.), can be achieved (Bannerot, Boulidard, Cauderon, and Tempe, 1974; Bannerot, Boulidard, and Chupeau, 1977), but at the expense of physiological vigour. At low, early spring

temperatures, such cms-rapeseed plants display a chlorosis that apparently
is the consequence of genomic incompatibility between the rapeseed nucleus
and the radish chloroplasts.

Pelletier, Primard, Vedel, Chetrit, Remy, Rousselle, and Bernard (1983)
corrected this defect by fusing protoplasts of normal, male fertile (mf)
B. napus with those of *cms B. napus* and then selecting from regenerants that
were both non-chlorotic under conditions of low temperature, as well as male
sterile. Later, Pelletier, Vedel and Belliard (1985) reported that chloroplast
genomes in the fused cells segregate during early post-fusion mitosis and
that the resulting cybrids possess either the one or the other parental
type of plastid genome. Parental mitochondrial genomes, on the other hand,
usually recombine, giving rise in the fusion products to new types of
mitochondrial DNA (see also chapters in this book by Vedel *et al.*, Pelletier,
and Barsby *et al.*).

That the manipulation of the chondriome is a potentially powerful tool
in modern plant breeding was demonstrated in 1983 in the brassicas by
Pelletier *et al.* with the construction of a three-way alloplasmic rapeseed
cybrid combining both cms radish cytoplasm and atrazine resistant birdseed
rape (*B. campestris*) cytoplasm, the latter encoded by chloroplast DNA.

In assembling the components for the production of hybrid rapeseed we
have attempted to obviate the unavoidable delay caused by selection and
usually also by backcrossing of progeny, which has to be painstakingly
selected from the fusion products of cms- and mf-*B. napus* protoplasts. This
contribution deals with some actual and potential strategies that may be
adopted *in vitro* for the production of cms-rapeseed plants with full
physiological vigour.

Terminology

Fractionation of a protoplast population using high speed density gradient
centrifugation may result in (1) non-fractionated protoplasts, (2) vacuoles
and vacuoplasts (vacuoles enclosed by a plasmalemma), (3) enucleated protoplasts or cytoplasts, (4) evacuolated as well as partially evacuolated
protoplasts, and (5) nucleoplasts (nuclei enclosed by a plasmalemma). In
the past, the term microplast has been applied to small, enucleated and
evacuolated or partially evacuolated protoplasts, and the term mini-
protoplast to small, nucleated but partially evacuolated protoplasts (Fig.1).

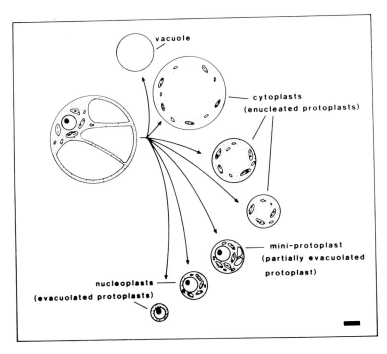

Fig. 1 - Enrichment of protoplasts and subprotoplasts can be achieved
by density gradient fractionation. See Introduction for
terminology. Bar represents 5 m.

Considerable variation in diameter and density exists among the sub-protoplast
components of a density gradient fractionation. In addition to debris in the
form of nuclei and chloroplasts, the pellet usually consists of mini-
protoplasts and nucleoplasts. The nucleoplasts of *Brassica napus* mesophyll
have diameters (arbitrarily chosen) of about 15 μm and smaller, and consist
of nuclei enclosed by a parietal layer of cytoplasm and a plasmalemma
(ca $\leq 10\mu$m) or by chloroplast-containing cytoplasm bounded by a plasmalemma (ca 10-15 μm). In this paper, the terms nucleoplast and mini-
protoplast may be used interchangeably.

    Abbreviations - ALA, 5-aminolevulinic acid; cms, cytoplasmically male
sterile or cytoplasmic male sterility; HEPES, hydroxyethylpiperazine-N-2-et-
hanesulfonic acid; MES, 2-(N-morpholino) ethanesulphonic acid; mf, male
fertile; PEG, polyethylene glycol; SAN 9789 or norflurazon, 4-chloro-5-
methylamino-2-(3-trifluoromethylphenyl)-pyridazin-3(2H)one.

STRATEGIES IN THE SYNTHEIS OF CMS-CYBRID *B. NAPUS* WITH NON-DEFECTIVE
PLASTID CHONDRIOMES

Following is a summary outline of the principles as well as the protoplast
manipulations applicable to the establishment of heteroplasmic cms-rapeseed
cybrids with restored physiological vigour.

A. Protoplast-protoplast or protoplast-cytoplast fusion

```
            cms-B. napus            mf-B. napus
            (2n=38, radish          (2n=38, rapeseed
            cytoplasm, defective)   cytoplasm, normal)

1.  protoplast x protoplast ──────────────    cms-B. napus─┐
                                              (2n=76)      │
                                                 x         │
                                              mf-B. napus ─┘
                                              (2n=38)      │
                                                           │
2.  protoplast x cytoplast ─────────────────────────────── cms-B. napus
                                                           (2n=38, cybrid,
3.  cytoplast  x protoplast ───────────────────────────── corrected)

4.  cytoplast  x protoplast ──────────────────────────────
    (minus          (or mini-
    chloro-         protoplast or nucleoplast)
    plasts)
5.  protoplast x cytoplast ──────────────────────────────┘
    (minus
    chloro-
    plasts)
```

B. Protoplast manipulation

1. Cytoplast formation by:

    1.1 elimination of nucleus by X-irradiation

    1.2 enrichment by discontinuous density-gradient centrifugation.

2. Mini-protoplast and nucleoplast formation by:

    2.1 enrichment by continuous density-gradient centrifugation

    2.2 enrichment by discontinuous density-gradient centrifugation.

C. Chloroplast elimination

1. Pathologic accumulation of protochlorophyllide followed by:

    1.1 irradiation, isolation, fusion and regeneration of protoplasts

    1.2 isolation, irradiation, fusion and regeneration of protoplasts

    1.3 isolation, fusion, irradiation and regeneration of protoplasts.

2. Photobleaching by irradiation after inhibition of carotenoid synthesis using norflurazon (SAN 9789) followed by:

    2.1 isolation, fusion and regeneration of protoplasts.

3. Selection by micromanipulation of a single cytoplast and a single protoplast followed by:

    3.1 elimination of the chloroplasts from the cytoplast using an helium-neon laser emitting at 632.8 nm with or without previous protochlorophyllide induction using ALA, followed by electrofusion and regeneration.

D. Screening of regenerated plantlings through:

    1. vernalization and observation of flower morphology

    2. restriction analysis of chloroplast and mitochondrial DNA

    3. low-temperature treatment and determination of variable chlorophyll fluorescence.

PROTOPLAST-PROTOPLAST OR PROTOPLAST-CYTOPLAST FUSION

Chondriome substitution is mediated by fusion, which can be induced by either polyethylene glycol (PEG) or an electric field. It is essential that at least one of the fusion partners represents a genotype that is regenerable and regeneration frequency is enhanced if at least one is of hypocotylary (Glimelius 1984) or petiolar origin.

Isolation and fusion procedures

Brassica napus was grown from either surface-sterilized seed, or axenically from shoot tip cultures in two-week rotation, under continuous light at 25 C with a photon fluence rate of 30 μmol m^{-2} s^{-1} (Osram Fluora 40 W fluorescent lamps).

Petioles of a cms line (R1466) and leaf blades of a normal line (R5903) were diced and incubated overnight in a solution consisting of 100 mM mannitol, 400 mM sucrose, 1mM CaCl$_2$-2H$_2$O and 5 mM MES, pH 5.5. The incubation solution included Macerozyme R-10 (0.1% for mesophyll and 0.3% for petiolar tissues) and Cellulase Onozuka R-10 (mesophyll-0.5%, petioles-1.5%). Both enzymes were obtained from Yakult Honsha Co. Ltd., Tokyo, Japan. Protoplasts were retrieved by flotation and washed twice in enzyme-free incubation medium. For enrichment of cytoplasts and mini-protoplasts, the washed protoplasts were transferred to respective overlayering media, which were 0.5 M with respect to mannitol (see Protoplast manipulation below).

Immediately prior to fusion with PEG, in a method modified by one of us (CJ) from Kao and Michayluk (1974) and Haydn, Lazar and Dudits (1977), protoplasts and cytoplasts were washed in a glucose rinse (400 mM glucose, 5 mM CaCl-2M O, 1 mM KH PO and 5 mM MES, pH 5.5), centrifuged at 100x g for 5 min, and the pellet resuspended and held at a density of 5 x 10^5 protoplasts ml^{-1} in the glucose rinse. For fusion, equal volumes of the two rapeseed populations were blended carefully and distributed as 50μ l droplets in petri dishes. After 15 min 50μ l 50% PEG (MW 1540, containing 100 mM glucose, 10 mM CaCl$_2$ 0.1 mM KH$_2$PO$_4$; pH 5.5) were added to each, and after a further 5 min, 2 ml of PEG dilution medium AB (A: 1 mM glycine, pH 10.5; B: 800 mM glucose, 100 mM CaCl$_2$. 2H$_2$O: 10% dimethylsulfoxide, pH 5.5: A and B mixed in equal volumes immediately before use were added slowly. The fused protoplasts were then washed twice with glucose rinse before addition of culture medium KM8p according to Glimelius (1984). In the case of electric field-induced fusion the blended populations were collected in a 1-ml hypodermic syringe and fusion carried out according to Zachrisson and Bornman (1984).

With these two approaches fusion frequencies of 5 to 10% for PEG- and as high as 20% for electric field-induced fusion have been obtained.

Rapeseed is an allopolyploid or amphidiploid (2n=38), but where petiole (or hyocotyl) protoplasts of the cms-line were fused with mesophyll protoplasts of the normal line according to strategy A1 (and Fig. 2A,B), we found 60% and more of the regenerated plants to be autopolyploid or amphitetraploid (2n=76). This predilection for synkaryony necessitates extensive backcrossing. In the remaining 40% of regenerants, sustained screening is required to identify the desired cybrid combination as shown in Fig. 2c.

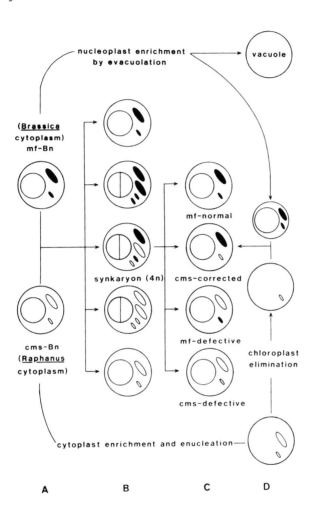

Fig. 2 - Fusion of male fertile (mf) - *Brassica napus* protoplasts with cytoplasmically male sterile (*cms*) - *B. napus* protoplasts containing *Raphanus sativus* cytoplasm with low temperature photosynthetic sensitivity (a), may give rise to a number of alternative products (b). However, more than 60% of such fusions may result in synkaryony, requiring backcrossing of such autopolyploid (amphi-tetraploid) plants with normal, amphidiploid *B. napus*.

> The backcrossed products (c) are screened for cms-B. napus
> plants with restored photosynthetic vigour. Evacuolation
> of mf-B. napus protoplasts results in mini- protoplasts or
> nucleoplasts and enucleation of cms-B napus protoplasts
> results in cytoplasts. Fusion of the evacuolated proto-
> plasts with the cytoplasts from which, in addition, the
> Raphanus chloroplasts have been eliminated (d), leads to
> the direct production of cms-corrected B. napus plants (c).

As indicated under strategies for synthesis of non-defective cms-cybrid
B. napus, one of the procedures (strategy A4; Fig. 2d) that avoids the
complications referred to above involves (1) the production of cms-B. napus
cytoplasts and (2) elimination of their chloroplasts together with (3) fusion
of these cytoplasts with mf-B. napus protoplasts (or mini-protoplasts or
nucleoplasts, Fig. 1).

PROTOPLAST MANIPULATION

Density gradient centrifugation is an effective substitute for and avoids
some of the problems associated with X-irradiation as a means of eliminating
the protoplasts' nuclear genome. In addition to enucleation, this technique
simultaneously offers the advantage of evacuolation of protoplasts.
As pointed out by Griesbach and Sink (1983), the presence of the vacuole
may have undesirable consequences in fusion. We therefore selected
strategies B1.2 and B2 to prepare enucleated protoplasts (cytoplasts) and
evacuolated protoplasts (mini-protoplasts, nucleoplasts), respectively.

Using cms-B. napus and mf-B. napus, we compared the methods of
Lorz, Paszkowski, Dirks-Ventling and Potrykus, (1981), Archer, Landgren,
and Bonnett (1982) and Griesbach and Sink (1983), who showed that with
density gradient centrifugation, protoplasts and cytoplasts can be enriched
through evacuolation and enucleation, respectively.

The low speed method reported by Archer et al., (1982) to enrich
cytoplasts from a protoplast population of Nicotiana debneyi, N. glutinosa
and N. glauca could not be repeated for B. napus. We found that with low
speed (200 xg) the best gradient was a four-step one of 50, 20, 10 and 5%
Percoll (Pharmacia, Uppsala, Sweden) in, and overlayed by 500 mM mannitol,
1 mM $CaCl_2-2H_2O$ and 5 mM MES, pH 5.5. Even with this gradient fractions of

proto- and subprotoplasts were too mixed and in contrast to Archer et al., nucleated protoplasts were not found at the interface of the overlayering medium and Percoll. This fraction consisted mainly of cytoplasts and large vacuolated protoplasts, and it was estimated that an enrichment of the cytoplast fraction by at most 5% could be attained. With our protoplast isolation method, cytoplasts comprise about 15 to 20% of the naked cell population isolated from the petioles of rapeseed.

Fractionation of protoplasts in discontinuous iso-osmotic Percoll density gradients at 20 000 to 40 000 xg was developed by Lorz et al., (1981). We obtained highly purified cytoplast fractions from B. napus petiolar tissue isolates using a three-step gradient of 50, 30 and 20% Percoll (Lorz et al., used 50, 20, 5, 0%) in a 500 mM mannitol medium as above, and overlayered with the same medium containing 10^6 protoplasts per 5 ml tube and centrifuged at 30 000 xg for 30 min at 15 C. The cytoplasts (variable diameter) fractionated distinctly within the 20% Percoll band, whereas the miniprotoplasts (diameter$\geqslant$15 μm) sedimented.

Griesbach and Sink (1983) reported evacuolation of Petunia parodii mesophyll protoplasts in a continuous Percoll gradient centrifuged at 150 000x g. Their procedure produced 100% evacuolation, gave 40% visibility and according to them allowed regeneration of plants. In our case, we first pre-centrifuged Percoll for 30 min at 150 000 xg, overlayed the gradient with 10^6 mf-B. napus mesophyll protoplasts suspended in 500 mM mannitol, 100 mM $CaCl_2$ $2H_2O$ and 5 mm HEPES, pH 7 (Griesbach and Sink 1983), and then centrifuged the suspension for 30 min at 150 000 xg at 15 C. Mini-protoplasts (diameter$\geqslant$15 μm) and nucleoplasts ($\geqslant$15 μm in diameter) were withdrawn from the surface of the silicone oil pellet and washed in 400 mM glucose rinse in preparation for fusion (see above).

The ability of nucleoplasts to hydrolyze fluorescein diacetate to fluorescein, often taken as an indication of viability (Bornman and Bornman 1985), was not affected by the procedures of density gradient centrifugation. When cultured separately we found the nucleoplasts generally incapable of division, suggesting that the critical ratio of nuclear to cytoplasmic volume had been exceeded. However, the fusion products of nucleoplasts and cytoplasts were capable of cell wall resynthesis and division, but at low frequency. Division frequencies were higher in the case of cytoplasts and non-fractionated protoplasts, but then inter-protoplast fusion (Fig. 2c)

made it more difficult to select cytoplast-protoplast fusion products
(Fig. 2d).

CHLOROPLAST ELIMINATION

In principle, plastids can be eliminated from protoplasts in three different
ways: removal, blocking of plastid DNA replication, or destruction. Removal
can involve either micromanipulation of individual protoplasts or electric
field-induced release of the chloroplasts (Zimmermann, Kuppers, and Salhani,
1982). Selective blocking of plastid replication has been achieved in the
green flagellate *Euglena* by application of ultraviolet radiation (Schiff,
Lyman and Epstein, 1961), high temperature, and application of chemical
inhibitors such as streptomycin and nalidixic acid (Ebringer 1972).

   We chose chloroplast destruction as a first approach, since removal is
unpractical and the blocking of plastid replication is best carried out on
cells in suspension culture. Furthermore, since cells of higher plants
probably do not survive long without functional plastids, there would be
little possibility of eliminating excess inhibitor before protoplast fusion.

Sensitisation of plastids by accumulation of protochlorophyllide and destruction by irradiation

Because plastids are the only cell organelles that contain large amounts of
pigments able to absorb long-wavelength light, they can be specifically
destroyed by such light. Very intense red light, absorbed by chlorophyll,
may destroy chloroplasts by the creation of triplet chlorophyll, which in
turn causes formation of singlet excited oxygen, which attacks various
membrane constituents. Plastids may be made more susceptible to light by
pretreatments which cause a pathologic accumulation of protochlorophyllide,
or by blocking carotenoid synthesis.

   There are two ways of inducing accumulation of protochlorophyllide.
Marschner (1964, 1965) and Marschner and Gunther (1966) detected that
barley leaves were easily killed by light if they were first permitted to take
up caesium ions. If a first illumination was followed by a period of darkness, large amounts of protochlorophyllide accumulated. A second irradiation
produced the lethal effect by causing light absorption in the *inactive*
accumulated protochlorophyllide. *Inactive* refers to protochlorophyllide
that is not bound to NADPH-protochlorophyllide-reductase, and thus cannot be

directly photoreduced to chlorophyllide a. Instead, the absorbed light energy takes a destructive path, which probably involves photodynamic action (Blum 1941) with formation of singlet excited oxygen and its subsequent attack on membrane constituents.

We had inconsistent responses with caesium ions on rapeseed seedlings and instead used 5-aminolevulinic acid (ALA), a precursor for the synthesis of porphyrins and chlorophyll. In the biosynthesis of chlorophyll there is a feedback loop limiting the synthesis of ALA:

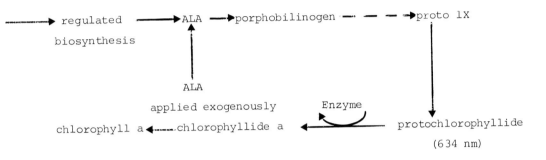

Protochlorophyllide is converted to chlorophyllide A by light acting on a protochlorophyllide-enzyme-NADPH complex. The resulting chlorophyllide a-enzyme-NADP complex dissociates, and the enzyme can combine with more protochlorophyllide. The concentration of free protochlorophyllide is thus decreased, resulting in a release from inhibition of ALA synthesis (Halliwell 1981). The rate of ALA synthesis is therefore always adjusted to the rate of chlorophyll formation and only traces of free protochlorophyllide, which absorbs maximally at 634 nm, are present.

If exogenous ALA is added to the system, large amounts of free protochlorophyllide are accumulated. Since it has to be enzyme bound to make possible the use of absorbed light energy for photoconversion to chlorophyllide a, light energy absorbed in free protochlorophyllide takes another course: probably triplet protochlorophyllide and then singlet excited oxygen are formed. In any case, the end result is destruction both of pigments and other plastid components upon illumination. Because free protochlorophyllide absorbs maximally at ca 634 nm, light of this wavelength is the most potent.

The method of plastid sensitization and destruction chosen was that corresponding to strategy Cl.2 and involved overnight enzyme incubation of petiolar tissues of *cms* rapeseed in the presence of 5 mM ALA, isolation of the protoplasts, enrichment of the cytoplast fraction, and subsequent irradiation of the cytoplasts with intense red light. Light from a 900 W xenon lamp was collimated with a glass lens and filtered through a Calflex C (Balzers, Liechtenstein) infrared reflecting interference filter, 20 mm of plexiglass, and a 2-mm RGl glass filter (Schott and GEn., Mainz). The spectral range was ca 600–800 nm and the irradiance either 80 W m^{-2} over ca 70 cm^2 or 420 W m^{-2} over ca 10 cm^2. Table I shows the decrease of total chlorophyll with time of irradiation in protoplasts isolated from petiolar tissues of *cms* rapeseed incubated for 12 hours in 5 mM ALA. After 14 h of irradiation, chlorophyll content had already been reduced by 70%.

Photobleaching by inhibition of carotenoid synthesis

When chlorophyll absorbs photons, the chlorophyll molecules are changed from the ground state to the first excited singlet state. Photosynthesis is driven by the energy released when the molecules return from the excited singlet state to the ground state:

Table I Time-course effects of irradiation on protoplasts isolated from cms-*Brassica napus* petiolar tissue exposed to 5 mM 5-aminolevulinic acid for 12 h.

Irradiation time, h	Decrease of total chlorophyll from dark control, %
0	12.3
0.5	20.6
1.0	37.6
2.0	48.4
3.0	56.3
4.0	69.1

A small fraction of the excited singlet molecules is not directly de-excited to the ground state; instead it is transferred to the excited triplet state. Triplet chlorophyll readily interacts with molecular oxygen, O_2, because oxygen is a triplet molecule in the ground state. This interaction results in ground state chlorophyll and singlet excited oxygen, $^1O^*_2$. Singlet excited oxygen is highly reactive and able to attack many compounds, including (ground state) chlorophyll. Chlorophyll is then oxygenated to $Chl.O_2$, a labile molecule which may decompose, resulting in the irreversible destruction of chlorophyll.

The main function of carotenoids in the chloroplast is the counteraction or prevention of the destruction of chlorophyll (Halliwell 1981). This protection appears to take place by two mechanisms:

(1) carotenoids de-excite triplet chlorophyll, before it reacts with oxygen; and (2) one or more of the carotenes (oxygen-free carotenoids) reduce chlorophyll peroxide under the formation of carotene epoxide. Carotene epoxide is enzymatically re-converted to carotene with NADPH as reductant.

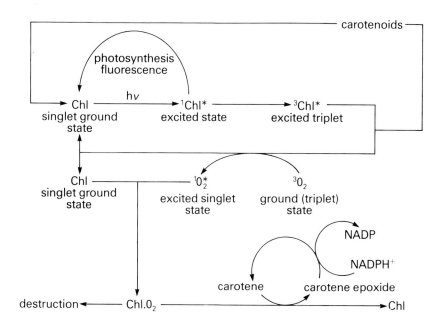

Even if only a small fraction of the excitation events results in the formation of triplet excited chlorophyll, excitations follow each other so rapidly that all the chlorophyll would be rapidly destroyed in ordinary daylight were it not for the protective action of carotenoids.

Norflurazon or SAN 9789 (4-chloro-5-methylamino-2-(3-trifluoromethylphenyl)-pyridazin-3(2H)one) inhibits carotenoid synthesis by blocking the conversion of the precursor phytoene to phytofluene (Schwartzbach and Schiff 1983). Phytoene, which then accumulates, is colourless (it absorbs only UV) and does not protect chlorophyll:

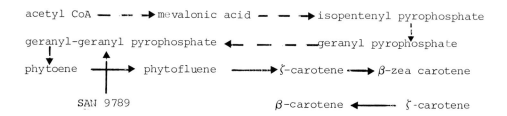

Following strategy C2, rapeseed seedlings turned white when germinated on 1% nutrient agar containing 3% sucrose and ca 100 g ml^{-1} SAN 9789, at 25°C and constant light in the 670-680 nm range. Protoplasts were isolated from the petioles of such seedlings, and cytoplasts were obtained by the enrichment procedures discussed earlier. The cytoplasts were then fused with either mf-B. $napus$ protoplasts or with mini-protoplasts or nucleoplasts.

Selective chloroplast destruction by helium-neon (He-Ne) laser microbeam

The beam of an He-Ne laser rated at 4 to 6 mW and mounted on an inverted microscope can be focused to a spot 5 to 8 μm in diameter, and is therefore especially suited for irradiation of single selected protoplasts. Because light of ca 634 nm wavelength is maximally absorbed by free protochlorophyllide, the He-Ne laser which emits at 632.8 nm is ideal from this point of view. Examination of leaf discs floated for 12 h on a 5 mM solution of ALA or on water, and irradiated with the laser, show a complete loss of chlorophyll within 10 (ALA) to 30 (water) min of treatment over the area corresponding to the diameter of the beam. Chloroplasts in individual cytoplasts from cms-rapeseed tissue exposed to ALA are quickly destroyed.

Although we still lack corroborating evidence, we believe that the method pioneered by Koop, Dirk, Wolff and Schweiger (1983) of somatically electrofusing two selected single protoplasts, can be made applicable to cybridisation. In fact, it is possible to regenerate callus from selected single protoplasts of B. $napus$. The problem that we confronted relates to post-fusion micro-manipulation of the cybrid cell. However, were it possible to regenerate a plant from the fusion product of a cms-rapeseed cytoplast (with laser-destroyed chloroplasts) and a normal rapeseed protoplast or mini-protoplast, it would eliminate virtually all post-fusion selection procedures.

SCREENING OF PLANTS REGENERATED FROM FUSION PRODUCTS

Rapeseed is a biennial. Vernalisation for seven weeks at 4-6 C under a 16-h photoperiod, photon fluence rate 15-18 μmol m^{-2} s^{-1}, of plants regenerated from the fusion products of cms-*B. napus* cytoplasts and normal *B. napus* protoplasts leads to flower induction. Flower morphology can therefore be used to confirm persistence of the cms-radish gene, whereas absence of chlorosis under the low temperature conditions during vernalization might be taken as a preliminary indication of the correction of the chlorophyll defect. Restriction analyses of organelle DNA can be used to verify both retention of cytoplasmic male sterility and transfer of the *B.napus* chloroplast genome. Where this kind of analysis cannot be carried out, variable chlorophyll fluorescence (strategy D3) under carefully controlled experimental conditions is an alternative parameter for monitoring the successful substitution of radish with rapeseed chloroplasts.

Variable chlorophyll fluorescence

A portion of the light intercepted by a plant is absorbed by the photosynthetic pigments and excites chlorophyll a. Under optimal conditions *ca* 85% of this energy is used in photosynthesis. The rest is lost as heat, or it is radiated as fluorescence. At physiological temperature, fluorescence emanates mostly from Chl a of photosystem II, with a main band maximum at 685 nm.

When a leaf is irradiated (with blue or red light), photosynthesis does not proceed immediately. On the way to an overall steady state, the photosynthetic apparatus passes through several transitory stages. These events are known as induction phenomena. This induction of Chl a fluorescence has been used as an intrinsic probe for changes in the photosynthetic electron transport, since the impairment of the photosynthetic system is reflected in the shape of the induction curves. So-called variable fluroescence is the parameter most often used: that is,

The chondriome and plant breeding

$$F_v - F_{max} - F_o$$

where F_{max} = maximum fluorescence, F_o = initial level of fluorescence, and F_v = variable fluroescence (Fig. 3).

Upon illumination, after a period of darkness, the fluorescence yield immediately reaches an initial level (F_o). This is followed by a rise from F_o to a maximum level (F_{max}). If the photosynthetic system is damaged, F_v is usually decreased.

Fluorescence measurements were made on the first fully-expanded leaf below the shoot apex of seedling plants of normal male fertile, cms-defective and cms-corrected B. napus, as well as of R. sativus, donor of the cms gene. Measurements were carried out using a modified Aminco-Bowman spectrophotofluorometer as described previously by Bornman, Bjorn and Akerlund (1984). Long wavelength fluorescence (> 710 nm) was measured. Plants were six weeks old, having been transferred at the commencement of week three to a growth cabinet at $10\,^{\circ}C$ under a 12-h photoperiod and photon fluence rate of 100 μmol m^{-2} s^{-1} (Sylvania, cool white, fluorescent, 115 W). At Durihy in Scandinavia, temperatures of $10\,^{\circ}C$ and lower are often experienced in conjunction with relatively high light intensities.

As shown in Table II, the mean F_v of R. sativus, mf-B. napus and cms-B. napus, the latter corrected according to strategies A4, B1.2, B2.1,

Table II   Relative values from fluorescence induction curves of
Raphanus satisvus, mf-Brassica napus, cms-B. napus with
defective nucleochloroplastic genome and cms-B.napus with
corrected nucleochloroplastic genome. F_o, initial level of
fluorescence; F_{max}, maximum fluorescence; F_v, variable
fluorescence. Culture temperature, $10\,^{\circ}C$.

Plant	F_{max}	F_o	F_v
cms-B.napus (defective)	90	35	55
R.sativus	59	17	42
mf-B.napus	53	13	39
cms-B.napus (corrected)	40	7	33

C2.1, D1, D3, Fig. 1b, is about 30% lower than that of cms-*B. napus* containing the defective radish chloroplast genome. The increased fluorescence displayed by the *cms-B. napus* plant with the radish chloroplast genome, in this case reflects a decreased ability of the use of chemical energy in photosynthesis.

CONCLUSION

It has been demonstrated that the chloroplast genome of one genus can be sensitized and selectively destroyed and, using fusion, somatically substituted with that of another genus. It was also shown that techniques for the fractionation of protoplasts can be used to enhance both the efficiency of cytoplasmic organelle substitution and the selectivity of the procedure. The production of cms-*B. napus* plants with restored physiological vigour was achieved, corroborating work reported earlier in France by G. Pelletier and co-workers.

REFERENCES

Archer, E.K., Landgren, C.R., and Bonnett, H.T. (1982). Cytoplast formation and enrichment from mesophyll tissues of *Nicotiana* spp. *Plant Sci. Lett. 25* : 175 - 85.

Bannerot, H., Boulidard, L., Cauderon, V., and Tempe, J. (1974). Transfer of cytoplasmic male sterility from *Raphanus sativus* to *Brassica oleracea. Proc. Eucarpia Meeting Cruciferae,* 52 - 4.

Bannerot, H., Boulidard, L., and Chupeau, Y. (1977). Unexpected difficulties met with the radish cytoplasm. *Eucarpia Cruciferae Newsletter,* 2 - 16.

Blum, H.F. (1941). *Photodynamic Action and Diseases caused by Light.* Reinhold, London.

Bornman, C.H., and Bornman, J.F. (1985). Plant protoplast visibility. In *The Physiological Properties of Plant Protoplasts* (ed. Pilet, P.E.) pp. 29 - 36. Springer-Verlag, Berlin, Heidelberg. ISBN 3 - 540 - 15017 - X.

Bornman, J.F., Bjorn, L.O., and Akerlund, H.E. (1984). Action spectrum for inhibition by ultraviolet radiation of photosystem II activity in spinach thylakoids. *Photobiochem. Photobiophys. 8* : 305 - 13.

Ebringer, L. (1972). Are plastids derived from prokaryotic micro-organisms? Action of antibiotics on chloroplasts of *Euglena gracilis*. *J. Gen. Microbiol. 71* : 35 - 52.

Glimelius, K. (1984). High growth rate and regeneration capacity of hypocotyl protoplasts in some Brassicaceae. *Physiol. Plant. 61* : 38 - 44.

Griesbach, R.J., and Sink, K.C. (1983). Evacuolation of mesophyll protoplasts. *Plant Sci. Lett. 30* : 297 - 301.

Halliwell, B. (1981). *Chloroplast Metabolism*. Clarendon Press, Oxford.

Haydn, Z., Lazar, G., and Dudits, D. (1977). Increased frequency of polyethylene glycol induced protoplast fusion by dimethylsulfoxide. *Plant Sci. Lett. 10* : 357 - 60.

Koop, H. U., Dirk, J., Wolff, D., and Schweiger, H.G. (1983). Somatic hybridization of two selected single cells. *Cell Biol. Int. Repts. 7* : 1123 - 8.

Lorz, H., Paszkowski, J., Dirks-Ventling, C., and Potrykus, I. (1981). Isolation and characterization of cytoplasts and miniprotoplasts derived from protoplasts of cultured cells. *Physiol. Plant. 53* : 385 - 91.

Marschner, H. (1964). Chlorophyllbildung und Blattschaden bei Gerste unter dem Einfluss von Cäsiumionen. *Flora 154* : 30 - 51.

Marschner, H. (1965). Anreicherung von Porphyrinen und Protochlorophyllid in Gerstensprossen unter dem Einfluss von Cäsium. *Flora 155* : 558 - 72.

Marschner, H., and Gunther, I. (1966). Veranderungen der Feinstruktur der Chloroplasten in Gerstensprossen unter dem Einfluss von Cäsium. *Flora 156* : 684 - 96.

Pelletier, G., Primard, C., Vedel, F., Chetrit, P., Remy, R., Rouselle, P., and Renard, M. (1983). Intergeneric cytoplasmic hybridization in Cruciferae by protoplast fusion. *Mol. Gen. Genet. 191* : 244 - 50.

Pelletier, G., Vedel, F., and Belliard, G. (1985). Cybrids in genetics and breeding. In *Proc. of the 1st Nordic Cell and Tissue Culture Symposium on Research, Breeding and Production of Crop Plants*, Frostavallen, Sweden, March 5 - 9, 1984 (eds. Bornman, C.H., Heneen, W.K., Jensen, C.J., and Lundqvist, A.). *Hereditas* Suppl. Vol. 3. 49 - 56.

Remy, R., and Ambard-Bretteville, F. (1983). Two dimensional analysis of chloroplast proteins from normal and cytoplasmic male sterile *Brassica napus.* *Theor. Appl. Genet. 64* : 249 - 53.

Schiff, J.A., Lyman, H., and Epstein, H.T. (1961). Studies of chloroplast development in *Euglena* II. Photoreversal of the U.V. inhibition of green colony formation. *Biochim. Biophys. Acta 50* : 310 - 18.

Schwartzbach, S.D., and Schiff, J.A. (1983). Control of plastogenesis in *Euglena* - in *Photomorphogenesis* (eds. Shropshire, Jr., W., and Mohr, H.) Vol. 16A pp. 312 - 35. Springer-Verlag, Berlin, Heidelberg.

Vedel, F., and Mathieu, C. (1983). Physical and gene mapping of chloroplast DNA from normal and cytoplasmic male sterile (radish cytoplasm) lines of *Brassica napus. Cuur. Genet. 7* : 13 - 20.

Zachrisson, A., and Bornman, C.H. (1984). Application of electric field fusion in plant tissue culture. *Physiol. Plant. 61* : 314 - 20.

Zimmermann, U., Kuppers, G., and Salhani, N. (1982). Electric field-induced release of chloroplasts from plant protoplasts. *Naturwissenschaften 69* : 451 - 2.

12   Organelle genetics of somatic hybrid-cybrid progeny in higher plants

A. Kumar and S. Cooper-Bland

ABSTRACT

A survey of the progress made in the field of somatic hybridisation in the last 10 years is presented, which clearly shows that this method has provided a novel means for studying organelle genetics in higher plants. Of significance, are the assignments of some cytoplasmically inherited traits to specific types of organelle (i.e. chloroplast genomes: sensitivity to antibiotics, herbicides and tentoxin, cytoplasmic albinoisms, and the large subunit of RuBPCase; mitochondial genome: abnormal flower morphology and the synthesis of new polypeptides in cms plants). Additionally, applications of this method for transferring the cytoplasmically inherited agronomically important traits in crop plants and for studying the behaviour of organellar genomes in plants somatic hybrids/cybrids and in animals somatic hybrids/cybrids are discussed.

INTRODUCTION

Somatic hybridisation by induced protoplast fusion, unlike sexual hybridisation by gametic fusion, brings together in a heterokaryon the nuclear and organellar (i.e. chloroplast and mitochondria) genomes from both parents, and thus provides a unique opportunity for creating novel genotypes in somatic hybrid*/cybrid+ progeny (Davey and Kumar, 1983; Gleba and Sytnik, 1984). This artificial method of hybridisation is novel in that the natural sexual hybridisation method, in most cases, combines in a zygote the nuclear genomes from both parents but only the cytoplasmic organellar genomes from the female parent (Sears, 1983). In this chapter, a review of the progress made in the last 10 years in the production of novel genetic variability at the cytoplasmic organelles level by somatic hybridisation is described and its uses in studies on organelle genetics and in plant breeding are discussed.

FOOTNOTE

Hybrid: Due to nuclear fusion, production of plants possessing any type of parental nuclear genomic hybridity ranging from total to partial gene transfer from either parent.

+*Cybrid*: In the absence of any nuclear hybridity, production of plants possessing only one or other parental nuclear genome together with any type of parental organellar (i.e. chloroplast/mitochondria) genomic hybridity ranging from both parental organelles co-existing in a common cytoplasm to the mixing of one chloroplast with the other mitochondria. Also, any new type of chloroplast or mitochondria, derived from DNA recombination between two parental chloroplast DNAs or between two parental mitochondrial DNAs. Additionally, the existence of the cytoplasmic organelles of one parent with the nucleus of the other parent in plants are called cybrid.

SOMATIC HYBRIDISATION IN PLANTS

The first somatic hybrid plants produced by protoplast fusion were between *Nicotiana glauca* and *N. langsdorffii* in 1972 by Carlson and his associates. Interestingly, biochemical analysis for identification of the parental chloroplast types in these somatic hybrid plants showed the presence of chloroplasts only from *N. glauca* (Kung *et al.*, 1975). Since the plants analysed were from the self-crossed progeny of an original somatic hybrid plant, initially derived from a single protoplast fusion event, it was not possible to conclude from these experiments whether biparental inheritance of the parental chloroplasts would be possible by somatic hybridisation. This unique facet of somatic hybridisation was first demonstrated by Gleba, Butenko and Sytnik (1975) in the intraspecific somatic hybrid/cybrid progeny produced between cytoplasmic albino and nuclear albino mutants of *N. tabacum*. These somatic hybrid/cybrid plants were novel in showing biparental inheritance of the cytoplasmic organelles: in *Nicotiana* the cytoplasmic organelles are only maternally inherited in sexual hybrid progeny (Sears, 1983). Since then somatic hybrid/cybrid calli and plants have been produced in over 70 instances between plant species of varying degrees of evolutionary divergence (Pelletier and Chupeau, 1984). Table I provides in chronological order the important

The chondriome and plant breeding

research on the organelle genetics of somatic hybrid/cybrid progeny
produced via protoplast fusion in higher plants.

Table I   Landmarks in the organelle genetics of somatic hybrid/cybrid
          plants (1975-1985)

1975   Kung *et al.* made the first attempts to analyse the fate of the
        cytoplasmic organelles (chloroplasts) in somatic hybrid plants
        of *N. glauca* and *N. langsdorffii*; they showed that the plants
        contain only chloroplasts from *N. glauca*.

1975   Gleba *et al.* demonstrated that parental chloroplasts can be
        biparentally inherited in intraspecific somatic hybrid progeny,
        using a cytoplasmic albino and a nuclear albino mutant lines of
        N. tabacum.   Their experiments also showed that the parental nuclear
        and chloroplast genomes can co-segregate or can segregate inde-
        pendently creating novel somatic hybrid/cybrid plants.

1977   Chen *et al.* demonstrated that parental chloroplasts can be bi-
        parentally inherited in interspecific somatic hybrid progeny of
        N. glauca and *N. langsdorffii,* but they randomly segregate to
        homogeneity for one or other parental chloroplast type in each
        somatic hybrid plant.

1378   Belliard *et al.* observed the transfer of cytoplasmic male sterility
        (*cms*) trait at interspecific level, following protoplast fusion
        between a fertile line of *N. tabacum* and a *cms* line of *N. tabacum*
        possessing *N. debneyi* cytoplasm.   They also showed a lack of
        correlation between the *cms* trait and the chloroplast genome.

1978   Gleba *et. al.* showed a co-segregation pattern for traits such as
        cytoplasmic albinism and the large subunit polypeptide (LS)
        (encoded by ct-DNA) of RuBPCase in somatic hybrid plants produced
        between a cytoplasmic albino mutant line of *N. tabacum* and a cms
        line of *N. tabacum* possessing *N. debneyi* cytoplasm.   They also
        showed the co-existence of both parental chloroplast types in a
        somatic hybrid plant and their stable transmission in the subse-
        quent sexual progeny.

1978   Melchers *et al.* observed the biparental inheritance of parental chloroplasts in intergeneric somatic hybrid progeny produced between sexually incompatible species; *Solanum tuberosum* and *Lycopersicon esculentum,* but each plant possessed only one or other parental chloroplast type.

1979   Belliard *et al.* observed that parental mitochondrial genomes can recombine their DNA and create a novel mitochondrial genome in somatic cybrid plants. They also demonstrated a correlation between the *cms* trait and the mitochondrial genome.

1980   Aviv *et al.* demonstrated that traits such as tentoxin sensitivity, LS of RuBPCase and ct-DNA restriction digest pattern co-segregate in somatic hybrid/cybrid progeny in *Nicotiana.*

1980   Aviv and Galun demonstrated restoration of fertility in a *cms* line via interspecific somatic hybridisation, using X-ray irradiation based on the 'donor-recipient' system.

1981   Menzcel *et al.* showed a definite correlation between streptomycin resistance and chloroplast genome using interspecific somatic hybridisation in *Nicotiana.*

1982   Galun *et al.* observed that parental nuclear, chloroplast and mitochondrial genomes can segregate independently in somatic hybrid/cybrid progeny in *Nicotiana.*

1982   Binding *et al.* demonstrated the transfer of atrazine resistance, a cytoplasmic encoded agronomically important trait, in *Solanum* species by interspecific somatic hybridisation.

1983   Pelletier *et al.* demonstrated the transfer of two cytoplasmic encoded agronomically important traits, *cms* and atrazine in *Brassica* species by intergeneric somatic hybridisation.

1984   Cseplo *et al.* showed that a sterile line in *Nicotiana* possessing cytoplasmic mutation conferring lincomycin resistance can be rescued in a fertile line by interspecific somatic hybridisation.

1985   Medgyesy *et al,* demonstrated that parental chloroplast genomes can recombine their DNA and create a novel chloroplast genome in a somatic hybrid plant.

The chondriome and plant breeding

A theoretical model

Based on the results described in Table I, the laws of genetic inheritance
of the parental nuclear, chloroplast and mitochondrial genomes in somatic
hybrid/cybrid progeny can be defined. Construction of a theoretical model
to highlight the diverse patterns of nuclear-organellar genomic combinations
that can apparently be created in somatic hybrid/cybrid progeny by somatic
hybridisation is presented in Figs. la and lb.

Laws of genetic inheritance of parental nuclear (N), chloroplast (C), and mitochondrial (N) genomes in somatic hybrid/cyrbid progeny

1. Firstly, parental nuclear genomes can stay together and can recombine
   their DNA to create a novel nuclear genome or can segregate independently
   in somatic hybrid/cybrid progeny, viz N^r, N_1, N_2.

2. Secondly, parental chloroplast genomes can stay together, can
   segregate independently or can recombine their DNA to create a novel
   chloroplast genome, viz. $C_1 + C_2$, C_1, C_2, C^r.

3. Thirdly, parental mitochondrial genomes can stay together, can segregate
   independently, or can recombine their DNA to create a novel mitochondrial
   genome, viz. $M_1 + M_2$, M_1, M_2, M^r.

4. Fourthly, parental nuclear, chloroplast, and mitochondrial genomes
   can co-segregate or can segregate independently, viz. $(N_2\ C_1 + C_2\ M_1)$,
   $(N^r\ C_1\ M_1 + M_2)$, $(N_1\ C^r\ M_2)$ etc.

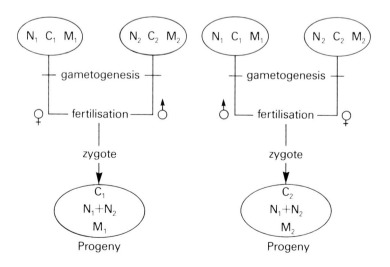

248

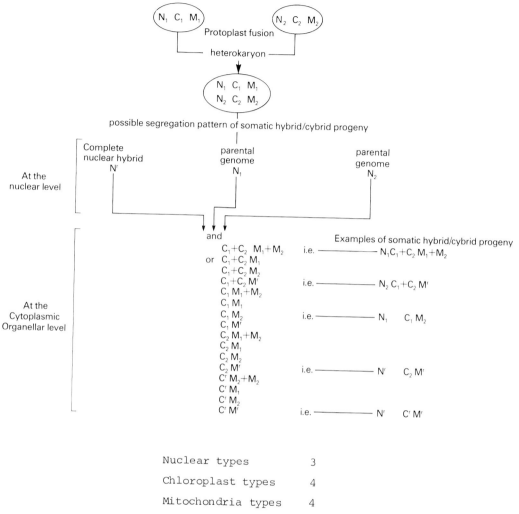

Nuclear types	3	
Chloroplast types	4	
Mitochondria types	4	

Therefore, a total of 3 x 4 x 4 = 48 different genotypic combinations of somatic hybrid/cybrid progeny can be created by somatic hybridisation (Fig. 1b).

A SURVEY OF THE AVAILABLE METHODS FOR CREATING GENETIC VARIABILITY AT THE CYTOPLASMIC ORGANELLES LEVEL IN PLANTS

Genetic variability at the cytoplasmic level in plants can be created in two different nuclear genetic backgrounds. Firstly, cytoplasmic genetic variability can be created in a hybrid nuclear background, i.e. $(N^r C_1 + C_2 M_1 + M_2)$ or $(N^r C_1 M_2)$ or $(N^r C^r M_1)$., etc., see for detail in Fig. 1b. Alternatively, the cytoplasmic genetic variability can be created in a chosen nuclear background, i.e. $(N_1 \, C_1 + C_2 \, M_1 + M_2)$ or $(N_1 \, C_1 \, M_2)$ or $(N_2 \, C_2 \, M^r)$ etc. (see Fig. 1b). These latter types of plant are called cybrids as their cytoplasmic organelles have been changed and not their nuclear backgrounds. Several methods have been devised in order to create genetic variability at the organelles level in higher plants which are described below.

Sexual hybridisation has been successfully used to produce cybrid plants between sexually compatible species. However, as previously mentioned, the uniparental maternal inheritance of cytoplasmic organelles which occurs in most cases during sexual hybridisation means that the only type of cybrid plants which can be produced are those possessing the nuclear genome of one parent and both the cytoplasmic organelles from the other parent $(N_1 \, C_2 \, M_2$ or $N_2 \, C_1 \, M_1)$. In order to produce these types of cybrid plants a lengthy recurrent backcrossing programme is required between the Fl hybrid and one of the parents to completely eliminate the nuclear genome of the cytoplasmic donor parent. Additionally, sexual hybridisation is also limited by the gene pool available to be utilized for cybrid production due to inherent sexual incompatibility phenomena operating between certain plant species.

Somatic hybridisation, by protoplast fusion, overcomes the barriers to sexual incompatibility and is potentially capable of producing a wide genotypic range of hybrid/cybrid plants (see Fig. 1b). Indeed, several studies have shown that it is possible to obtain a novel combination of chloroplast and mitochondria in somatic hybrid/cybrid plants (Pelletier and Chupeau, 1984; Kumar and Cocking, 1986). Essentially, the novel genotypic combinations are produced by the processes of fusion, somatic segregation and recombination between the parental nuclear, chloroplast and mitochondrial genomes in the regenerating hybrid cells (see Fig. 1b). Nevertheless, it should be noted that the majority of plants obtained, via straightforward protoplast fusion, are somatic nuclear hybrids rather than somatic cybrids (Galun and Aviv, 1983). The somatic nuclear hybrid plants possessing a novel cytoplasmic organellar genomic combination $(N^r \, C_1 \, M_2)$ are interesting biological materials in their own right and should be evaluated for their use in plant breeding. However, ideally one requires cybrid plants possessing a chosen nuclear genome and alien chloroplast and/or mitochondrial genomes, either for agronomic improvement, or for studying genomic interactions between nuclear-organellar genomes.

In order to produce efficiently such cybrid plants a modified experimental approach is required. The "donor-recipient" protoplast fusion technique developed by Zelcer, Aviv and Galun (1978), and further modified by Sidorov, Menczel, Nagy and Maliga (1980), fulfils this requirement and provides an excellent opportunity to study nuclear-organellar and organellar-organellar genomic interactions in heterokaryotic cells, and in any somatic cybrid plants (Galun and Aviv, 1983). This system is based on X-ray or γ-ray irradiation of the donor protoplasts, in order to debilitate their nuclear genomes, followed by fusion with competent recipient protoplasts. A similar result can be achieved but with greater experimental difficulty, using protoplast systems lacking nuclei (i.e. cytoplasts, sub-protoplasts) as a donor system (Maliga, Lorz, Lazar and Nagy, 1982; Davey and Kumar, 1983). Using this "donor-recipient" system, numerous experiments have been performed in order to determine the degree of organellar heterozygosity which can be created as a result of protoplast fusion between plant species with varying degrees of evolutionary diversity (Galun and Aviv, 1983; Menczel and Lorz, in Vasil, 1984).

The above-mentioned approaches can be limited by the transfer of organelles collectively (i.e., chloroplasts and mitochondria together) from one parent to the other. As both the chloroplasts and mitochondria have their own genomes which control independently specific physiological and metabolic characteristics, it would be ideal to transfer these organelles independently. Several methods have been investigated in order to facilitate transfer of chloroplasts or mitochondria independently using enriched population of the organelles. Uptake of isolated chloroplasts has been achieved by using protoplast fusogenic agents such as PEG, poly-L-ornithine and high pH Ca^{++} (Wallin, in Vasil, 1984). However, there are no convincing reports of the successful transfer of physiologically competent chloroplasts by this method. It has been suggested that the chloroplasts incur lethal damage during the process of uptake into the protoplast due either to their membrane fusion with the protoplast membrane or due to the damaging effect of fusogens (Davey and Kumar, 1983). Liposomes have been used in circumvent chloroplast membrane damage during the uptake or fusion process by encapsulating them in bilayered lipid vesicles (Giles, 1983). However, this has yet to be proven successful. A more direct approach could be the direct injection of an enriched chloroplast fraction into the cytoplasm of recipient protoplasts. One of the authors (A.K.) in collaboration with P. Jordan and D. Raineri at Southampton University has attempted to develop a system of direct microinjection using a 1 μm diameter glass micropipette and a Leitz micromanipulator system. However, to date our preliminary experiments have failed to show successful transfer of chloroplasts into protoplasts. One of the major hurdles in using isolated chloroplasts or mitochondria for either uptake or liposome or microinjection experiments is the problem of devising a protocol for their isolation in a physiological active state and as a clean enriched population together with devising medium capable of sustaining their viability. Other problems can be related to the polyploid nature of the organelle genomes in higher plants. Nevertheless, all the direct transfer methods mentioned above should be continued to be investigated as they offer a more specific transfer or cytoplasmic organelles into recipient protoplasts.

gated as they offer a more specific transfer of cytoplasmic organelles into principle or utilizing selectable phenotypic markers remain for the time being the most successful and practical methods for creating genetic variability at the organelle level in higher plants.

FATE OF THE CYTOPLASMIC ORGANELLES IN SOMATIC HYBRID/CYBRID PROGENY

Protoplast fusion combines nuclear, mitochondrial and chloroplast genomes of both fusion partners. An enforced co-existence of these three genetic systems of two (or more) different genotypic parents in a heterokaryon predictably evokes complex somatic incompatibility reactions due to inter-nuclear, inter-organellar and nucleo-organellar genomic interactions. The fate of parental cytoplasmic organelles in the somatic hybrid/cybrid plants regenerated from such heterokaryons will therefore largely depend on the extent of genomic divergence existing between the fusion partners, since this is likely to influence the replication potential of the chloroplast and mitochondrial genome in the cultured fusion products. Other factors, which can influence the fate of parental cytoplasmic organelles in somatic hybrid/cybrid plants are the physiological conditions of the parental protoplasts prior to their fusion (i.e. the number of chloroplasts and mitochondria per protoplast in each of the fusion partners, the mean number of genomes in each chloroplast and in each mitochondrion at the time of fusion), the nature of the fusogen employed (i.e. PEG or high pH Ca^{++}) and the nature of the selection system employed (i.e. albino complementation or antibiotic resistance) to select hybrid/cybrid cells from amongst the unfused parental and homokaryotic cells.

During the last 10 years, several biochemical and molecular markers have been assigned to the chloroplast genome (ct-DNA) e.g. cytoplasmic albino mutants, the large subunit of RuBPCase, tentoxin, antibiotic and herbicide sensitivity and the restriction digest pattern of (ct-DNA) and to the mitochondrial genome (mt-DNA), e.g. flower morphology of cytoplasmic male sterile (cms) plants, the synthesis of cms specific polypeptides and the restriction digest pattern of mt-DNA and these markers have been used

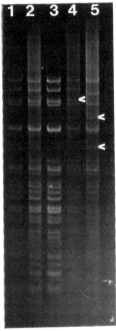

Fig. 2 - Restriction endonucleases fragment pattern analysis of ct-DNA in interspecific somatic hybrids of *Petunia*. *EcoRI* digests of ct-DNA were separated by 0.9% agarose slab gel electrophoresis, and the DNA fragments were visualised by staining with ethidium bromide. (1) *P. inflata*, (2) *P. hybrida*, (3) *P. parodii*, (4) *P. parodii* (+) *P. parviflora*, (5) *P. parviflora*. The arrow indicates the unique *EcoRI* fragments present in the ct-DNA of *P. parviflora* (<) and in that of *P. parodii*.

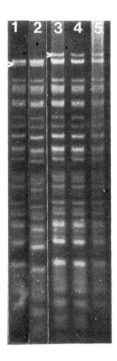

Fig. 3 - *Hpa II* digest pattern of (1) *P. hybrida*, (2) *P. inflata*
(3) *P. parodii*, (4) *P. parodii* (+) *P. inflata* and (5)
P. parodii (+) *P. hybrida*. The arrow indicates–additional
fragment in the ct-DNA digest pattern of *P. parodii* (**>**),
P. hybrida (**>**) and *P. inflata*.

extensively to study the fate of cytoplasmic organelles in resultant somatic
hybrid/cybrid plants (Davey and Kumar, 1983; Fluhr, 1983). These studies
have clearly demonstrated that the cytoplasmic organelles are both uni-
and/or bi-parentally inherited in the somatic hybrid/cybrid progeny, and
that a new type of chloroplast genome or mitochondrial genome can be
created in somatic hybrids/cybrids.

Fate of parental chloroplasts
Although the fusion products initially contain a mixed population of
parental chloroplasts, the somatic hybrid/cybrid plants subsequently recovered have been found, in most cases, to possess only one, or other,
parental chloroplast types (Chen, Wildman and Smith, 1977; Flick and
Evans, 1982; Fluhr, 1983). For instance, ct-DNA analysis of the three

different interspecific somatic hybrids in *Petunia*; *P. parodii* (+)
P. hybrida; *P. parodii* (+) *P. inflata* and *P. parodii* (+) *P. parviflora*,
(Figs. 2 and 3) show that only chloroplasts of *P. parodii* are present
in these somatic hybrids. The unidirectional segregation of the parental
chloroplast types in these three cases seems to be the result of the use
of a stringent selection pressure against one of the parents (Table III).
P. parodii was a common parent and used as a wild type for interspecific
somatic hybridisation with the three other albino mutant parents viz.
P. hybrida, *P. inflata* and *P. parviflora* (Kumar *et al.*, 1981, 1982).
By contrast, based on the analysis of isoelectric focusing pattern of
the large subunit polypeptides composition of RuBPCase of somatic hybrid
plants between a nitrate reductase deficient mutant line of *N. tabacum*
(NR) (+) a wild type line of *N. glutinosa* (Fig. 4) it was found that
both parental chloroplast types were present in the somatic hybrid plant
population but in each somatic hybrid plant there was only one type of
parental chloroplast (Table II). This result has been confirmed by more
recent tentoxin analysis (Cooper-Bland, Kumar, Pirrie, Pental and
Cocking, 1986).

In other experiments, individual plants with a mixture of chloroplast
types have been detected amongst the regenerated intra- and interspecific
somatic hybrid/cybrid plants (Gleba *et al.*, 1975; Fluhr, 1983, 1985;
Aviv *et al.*, 1984). There are two explanations for the co-existence of the
chloroplast types in cells of somatic hybrid/cybrid plants. Either, it may
relate to parental chloroplast genomic similarity or it may be because no
selection pressure was applied to the heterokaryotic cells during their
regeneration into plants. Recently, Gleba *et al.* (1985) have shown that
heterozygosity for parental chloroplast types is relatively stable in
hybrid/cybrid plants, and can be maintained even in their sexual progeny.
Obviously, the co-existence of parental chloroplast types in a common
cytoplasm would provide a unique opportunity for DNA recombination between
the two parental chloroplast genomes. However, until recently all
experimental protocols had failed to achieve this goal (Fluhr *et al.*,
1984) possibly due to the lack of a stringent selection system, or alternatively, due to the very low frequency at which chloroplast DNA recombination might occur. However, Maliga and his associates have been able recently
to devise a protocol based on the fusion of two different antibiotic

Table II   The fate of cytoplasmic organelles in somatic hybrid progeny

Fusion partners	Selection
1. *Petunia parodii* (WT) (+)*P. hybrida* (actinomycin D sensitive)	Media selection based on differential growth
2. *P. parodii* (WT) (+)*P. inflata* (cytoplasmic albino mutant) 3. *P. parodii* (WT) (+)*P. parviflora* (nuclear albino mutant)	Media and visual selection of green colonies, resulting after albino complementation

4. *Nicotiana tabacum* (NR⁻)
(+)*N. glutinosa* (WT)
 	
    Clone A - 1
    Clone A - 2
    Clone A - 7
    Clone A -10
	Complementation for
	active NR enzyme
	and visual selection of
	green, NR proficient
	colonies

NA - not analysed   NR⁻ - nitrate reductase deficient mutant
WT - wild type   M   - markers not available
LS - large subunits   SS   - small subunits

	Genetic markers					
Chromo- some No.	SS of Fl protein	LS of Fl protein	Tentoxin	Chloro- plast DNA	Mito- chondrial DNA	Fertility
28	M	M	M	*P. paro.*	NA	Fertile
28	M	M	M	*P. paro.*	NA	Fertile
31	*P. paro.*+ *P. parvi.*	M	M	*P. paro.*	NA	Sterile
69	*N. tab.*	*N. tab.*	*N. tab.*	NA	NA	Self-sterile
–	+	*N. tab.*	*N. tab.*	NA	NA	but cross-
68	*N. glut.*	*N. tab.*	*N. tab.*	NA	NA	fertile
7o		*N. glut.*	*N. glut.*	NA	NA	

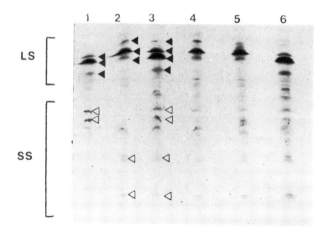

Fig. 4 - Subunit polypeptide composition of RuBPCase isolated
from somatic hybrid clones and their parents; *Nicotiana tabacum*
var. Gatensleben and *N. glutinosa*. Polypeptides of Fraction I
protein were separated by isoelectric focussing in 6.5% polyacrylamide slab gel containing 4% ampholine, pH 5-7 and 8M urea.
Three large subunit polypeptides (LS) were resolved in the pH 6
region of the electrofocusing gradient and the small subunit
(SS) polypeptides were resolved in the pH 5 region.   Polypeptide
bands were visualised by staining with 0.1% brilliant blue R-250.
Lane (1) *N. glutinosa*, (2) *N. tabacum*, (3) somatic hybrid (SH)
clone 1, (4) SH clone 2, (5) SH clone 7, (6) SH clone 10.

resistant cytoplasmic mutant lines, one lincomycin resistant and the other
streptomycin resistant (this latter line also possess a lesion in the
chloroplast DNA affecting chlorophyll synthesis). With the help of this
stringent selection system and a rigorous analysis of the somatic hybrid
progeny using recombinant DNA techniques, they were able to identify
chloroplast DNA recombination events in one somatic hybrid plant (Medgyesy,
Fejes and Maliga, 1985). The finding of chloroplast recombinants in
somatic hybrids suggests that chloroplast DNA recombination can be used to
create genetic variations in the chloroplast genomes and also to map
the genes on the chloroplast genome of higher plants just as has been very
successfully used in *Chlamydomonas* (Lemieux, Tormel, Seligny and Lee, 1984).

It is perhaps pertinent, and intriguing, to point out that DNA recombination between parental chloroplast genomes does not occur in species such as *Oenothera* (Chiu and Sears, 1985) and *Pelargonium* (Metzlaff, Borner and Hagemann, 1981), which show biparental inheritance of the organelles in sexual hybridisation.

Fate of the parental mitochondria

The mitochondria of the parental fusion partners are also initially mixed together in the cytoplasm of heterokaryons following protoplast fusion. However, the ensuing segregation pattern of the parental mitochondria differs from that of chloroplasts since mitochondria do not normally sort out randomly to homogeneity in the somatic hybrid/cybrid plants regenerated from the heterokaryons. There are only a few reports of such sorting of parental mitochondria in the somatic cybrid plants (Aviv *et al.*, 1984; Galun *et al.*, 1982). In most cases, parental mitochondrial DNA undergoes dramatic changes following protoplast fusion. The contrast between the behaviour of chloroplasts and mitochondria may be related to the structural organisation at the membrane level of these cytoplasmic organelles and also may be due to differences in their genomic organisation (Chourey *et al.*; Vedel *et al.*, in this volume). Analysis of most somatic hybrid/cybrid plants in *Nicotiana* indicate that the restriction digest patterns of mt-DNA are unique compared with the parental mt-DNA (Belliard *et al.*, 1979; Fluhr, 1983). Similar results have also recently been reported for *Petunia* (Clark *et al.*, 1985; Hanson and Conde, 1985), *Brassica* (Pelletier *et al.*, 1983; Chetrit *et al.*, 1985) and for *Solanum* (R. Kemble, personal communication). The unique mt-DNA restriction digest pattern in somatic hybrids/cybrids has been suggested to be a result of parental mt-DNA recombination (Belliard *et al.*, 1979; Hanson and Conde, 1985). This suggestion is based on DNA-DNA hybridisation results showing that the mitochondrial genome of somatic hybrid/cybrid consists of DNA fragments derived from both parents in novel combinations (Belliard *et al.*, 1985; Nagy *et al.*, 1983; Chetrit *et al.*, 1985; Rothenberg *et al.*, 1985). The possibility that intramolecular recombination in the mitochondrial genome might occur during *in vitro* culture of hybrid/cybrid cells and that this could be responsible for the mt-DNA variation observed in somatic hybrid/cybrid progeny has also been considered (Hanson and Conde, 1985; R. Kemble, personal communication).

The chondriome and plant breeding

It should be noted that *in vitro* culture has been found to induce mt-DNA rearrangements in maize (Kemble *et al.*, 1982), and in potato (Kemble and Shepard, 1984). However, this is apparently not the case in *Nicotiana* (Nagy *et al.*, 1983), or in *Petunia* (Hanson and Conde, 1985), as their mitochondrial DNAs appear quite stable under *in vitro* culture conditions. It is relevant to mention here that the *in vitro* culture process does not appear to induce any internal DNA rearrangement in the chloroplast genome (Kemble and Shepard, 1984; Hanson and Conde, 1985).

In summary, it can be concluded that following somatic hybridisation the parental chloroplasts and mitochondria of different plant species can co-exist together or segregate to homogeneity (parental chloroplasts and mitochondria segregating independently of each other) or can undergo recombination between different parental chloroplast or mitochondrial types in the somatic hybrid/cybrid progeny. This wide range of genetic variability of the cytoplast organelles is only possible through somatic hybridisation by protoplast fusion in higher plants. This novel facet of somatic hybridisation is now being used to characterise cytoplasmically inherited traits and to create useful novel genetic variability at the level of the organelle genomes in agronomically important crop species (see Pelletier *et al.* and Barsby *et al.* in this volume).

CHARACTERISATION OF CYTOPLASMICALLY INHERITED TRAITS VIA SOMATIC HYBRI-DISATION

Sexual hybridisation has been extensively used to assign specific traits to the nuclear genome but less extensively to the cytoplasmic organellar genomes due to the uniparental maternal inheritance of organelles during sexual hybridisation between most higher plant species. With the advent of somatic hybridisation it is now possible to study the patterns of inheritance of the cytoplasmically inherited traits in higher plants. In addition, the study of organelle genetics has greatly benefited from the advances in molecular techniques in recent years (Fluhr, 1983; Davey and Kumar, 1983). Currently, somatic hybridisation coupled with the biochemical and molecular analysis is being used to determine the genetic nature of cytoplasmically inherited traits such as resistance to antibiotics, to certain herbicides and to tentoxin and to obtain a better understanding of cytoplasmic albino mutants, cytoplasmic male sterility and of the large

260

subunit of RuBPCase. A summary of the progress made towards the characterisation of these traits is provided in Table III.

Table III Characterisation of the cytoplasmically inherited traits
via somatic hybridisation

I. Co-segregation of traits with the chloroplast genome

---

1. Atrazine sensitivity

2. Chloroplast DNA restriction digest patterns

3. Cytoplasmic albinism

4. Lincomycin sensitivity

5. Large subunit polypeptides composition of RuBPCase

6. Streptomycin sensitivity

7. Tentoxin sensitivity

II. Co-segregation of traits with the mitochondrial genome

---

1. Abnormal flower morphology in cytoplasmic male sterile
(cms) plants

2. Mitochondrial DNA restriction digest patterns

3. Polypeptides synthesis specific to cms lines

Modified from Gleba and Sytnik, 1984 and summarised from data in:
Gleba *et al.*, 1978; Aviv *et al.*, 1980; Menczel et al., 1981; Cseplo
et al., 1984; Belliard *et al.*, 1979; Charbonnier *et al.*, 1984; Gressel
et al., 1983; Pelletier *et al.*, 1983; Chetrit *et al.*, 1984; Cooper-
Bland *et al.*, 1986; Clark *et al.*, 1985.

SOME COMPARISONS OF THE ORGANELLE GENETICS OF SOMATIC HYBRIDS/CYBRIDS
IN PLANTS AND IN ANIMALS

Animal cells possess only two types of genetic system, i.e. nuclear and
mitochondrial genomes, whereas plant cells each possess three types of
genetic system, i.e. nuclear, mitochondrial and chloroplast genomes.
Therefore with regard to the behaviour of cytoplasmic organelles only a

Table IV   A comparison of organelle genetics of somatic hybrids/cybrids
in plants and in animals

Species	Fate of ct-DNA			
	C_1	C_2	$C_1 + C_2$	C^{r*}
Plants				
1. *Brassica*	+	+	−	−
2. *Daucus*	+	+	−	−
3. *Lycopersicon*	+	+	+	−
4. *Nicotiana*	+	+	+	+
5. *Petunia*	†	+	−	−
6. *Solanum*	+	+	−	−
Animals				
1. Mouse (+) Rat		CA		
2. Mouse (+) Chinese hamster		"		
3. Mouse (+) Human		"		
4. Rat (+) Human		"		

*r - intra or inter molecular DNA recombination
+, - observed
M = mitochondria

comparison of the behaviour of the mitochondrial genome in somatic hybrids/
cybrids of plants and animals can be made. The inheritance of the parental
mitochondrial genome is predominantly uniparental and maternal during
sexual hybridisation in animals (Gyllensten *et al.*, 1985) which is the
situation found during sexual hybridisation between most plant species

Fate of mt-DNA				References
M_1	M_2	$M_1 {}^+ M_2$	M^{r*}	
−	−	+[1]	+	Pelletier et al., 1983, Chetrit et al. 1984.
−	−	+	+	Matthews et al., 1985.
+	+	−	−	O'Connell and Hanson, 1985.
+	+	+	+	Belliard et al., 1978, 1979; Galun et al., 1982; Aviv et al., 1984; Gleba et al., 1985; Nagy et al., 1983; Medgyesy et al., 1985; Fluhr et al., 1983, 1984.
−	−	+	+	Kumar et al., 1982; Clark et al., 1983; Rothenberg et al., 1985.
+	+	+	+	Gressel et al., 1983; Kemble et al., 1985 (Per. comm.)
+	+	+	+	Hayashi et al., 1981, 1982, 1985.
+	+	−	−	De Francesco et al., 1983; Zuckerman et al., 1984; Solus and Eisenstadt, 1984.
+	+	+	+	Clayton et al., 1971; Attardi and Attardi, 1972; Coon et al., 1973; Horak et al., 1974.
+	+	+	+	De Francesco et al., 1980.

−, − not observed
CA = chloroplast absent
1 or 2 = parental type

(Sears, 1983). Therefore, in order to achieve biparental inheritance of the parental mitochondrial genomes in plants and in animals somatic hybridisation, by plant protoplast fusion or by animal cell fusion, respectively, is essential.

The chondriome and plant breeding

It is interesting to compare the fate of the parental mitochondrial genomes in animal somatic hybrid/cybrid progeny with the fate of mitochondrial genomes in plant somatic hybrid/cybrid progeny (Table IV). In somatic hybrids between mouse (+) human (De Francesco *et al.*, 1980) and between mouse (+) chinese hamster (Zuckerman *et al.*, 1984) that have been analysed, generally only one type of parental mitochondrial genome has been observed (Table IV). In these somatic hybrids, the absence of mitochondria of one parental species has been shown to be correlated with the loss of the corresponding parental chromosomes from the hybrid cell lines (De Francesco *et al.*, 1980). In other words, the presence of a single parental mitochondrial type in somatic hybrid cells is determined by the parent whose nuclear chromosomes are most stable. The possibility that the propagation of mt-DNA of one parental species in the hybrid cell depends on a particular chromosome or set of chromosomes has also been suggested (De Francesco *et al.*, 1980). Those working with plant somatic hybrids/cybrids should also be aware of this possibility, as it is quite common in heterokaryotic cells, especially produced between more distantly related species, to exhibit unidirectional loss of chromosome.

Concurrently, several reports have also demonstrated parental mt-DNA recombination in animal somatic hybrids/cybrids of human (+) mouse (Clayton *et al.*, 1971; Attardi and Attardi, 1972; Coon *et al.*, 1973; Horak *et al.*, 1974), human (+) rat (De Francesco *et al.*, 1980), and mouse (+) rat (Hayashi *et al.*, 1981)(Table 4). However, it should be noted that the observed high frequency of alteration in mt-DNA due to parental mt-DNA recombination or due to *in vitro* culture induced mutations in plant somatic hybrids/cybrids does not occur as frequently in animal somatic hybrids/cybrids (Hayashi *et al.*, 1982, 1985). It is not known why the plant mitochondrial genome behaves differently to the animal mitochondrial genome in somatic hybrids/cybrids. One of the factors that has been suggested to be responsible is the nature of the genomic construction of the mt-DNA itself. Plant mitochondrial genomes are very large, vary enormously in their size between species (250 to 2,500 kb), are labile and contain redundant DNA (Levings, 1983). In constrast, animal mitochondrial genomes are relatively small, are conservative in their size between species (15 to 18 kb), are not labile and do not contain redundant DNA (Levings, 1983). It would be interesting to study the fate of mitochondrial DNAs in somatic hybrids/cybrids between animal and plant since such novel somatic hybrids/cybrids can now be produced (Davey and Kumar, 1983). It is pertinent to point out here that in animal somatic hybrids/cybrids the fate of the parental cytoplasmic organelles can only be studied in the undifferentiated cell lines, since animal cells are not totipotent. However, in plant somatic hybrids/cybrids, the cytoplasmic organelles can be studied at all levels; from the heterokaryotic cells through to fully differentiated mature plants. Indeed, it has been observed in plant somatic hybrids that the differentiation process might determine the fate of the parental chloroplasts segregation patterns (see Chapter 14).

CONCLUSION

From this review, it is evident that somatic hybridisation by protoplast fusion allows biparental inheritance of both the nuclear and organellar genomes and also overcomes the sexual incompatibility barriers between plant species. It thereby provides a unique method for creating novel genotypes and for studying organelle genetics in higher plants. It is also becoming clear that the ensuing processes of somatic segregation, mitotic recombination and somaclonal variation acting on the nuclear and organellar genomes of the developing hybrid/cybrid cells can generate a wide range of genotypic combinations in subsequently regenerated somatic hybrid/cybrid plants (Kumar and Cocking, 1986). Somatic hybridisation, together with biochemical and molecular analysis, has already been successful in assigning certain cytoplasmically inherited traits on the organelle genomes (Fluhr et al., 1985). This method has also proved useful for transferring some cytoplasmically inherited agronomically important traits from wild species into crop species (Pelletier et al., this volume). A recent report describing chloroplast transformation using a Ti plasmid-derived vector system is encouraging (de Block et al., 1985) and this method may provide another novel means for creating novel genetic variability at the cytoplasmic organelles level and for studying organelle genetics in higher plants. Obviously, the eventually limited protoplast culture technology must be extended to a wider range of plant species, in particular to the crop species, in order to exploit more fully these novel methods.

The chondriome and plant breeding

ACKNOWLEDGEMENTS

Somatic hybridisation work in *Nicotiana* mentioned in this paper was funded by SERC to SC-B for her Ph.D. (at Nottingham University, Nottingham) with Professor E.C. Cocking; SC-B thanks Professor Cocking for his encouragement. Microinjection work was funded by AFRC to Dr. P. Jordan for a post-doctoral research fellow position held by AK in the Department of Biochemistry, Southampton University. AK thanks Professor M. Akhter, for his encouragement, and to Miss D. Raineri, Mr. S. Thomas and Dr. I. Cottingham for their helpful suggestions. We also thank Dr. A.C. Cuming (Leeds University) for critically reading the manuscript.

REFERENCES

Attardi, B., and Attardi, G. (1972). Fate of mitochondrial DNA in human-mouse somatic cell hybrids. *Proc. Natl. Acad. Sci. USA* 69 : 129 - 133.

Aviv, D., Arzee-Gonen, P., Bleichman, S., and Galun, E. (1984). Novel alloplasmic *Nicotiana* plants by "donor-recipient' protoplast fusion: Cybrid having *N. tabacum* or *N. sylvestris* nuclear genomes and either or both plastomes and chondriomes from alien species. *Mol. Gen. Genet.* 196 : 244 - 253.

Aviv, D., Fluhr, R., Edelman, M., and Galun, E. (1980). Progeny analysis of the interspecific somatic hybrids: *Nicotiana tabacum* (cms) + *N. sylvestris* with respect to nuclear and chloroplast markers. *Theor. Appl. Genet.*, 56 : 145 - 150.

Aviv, D., and Galun, E. (1980). Restoration of fertility in cytoplasmic male sterile (cms) *Nicotiana sylvestris* by fusion with X-ray irradiated *N. tabacum* protoplasts. *Theor. Appl. Genet.* 58 : 121 - 127.

Belliard, G., Pelletier, G., Vedel, F., and Quetier, F. (1978). Morphological characteristics and chloroplast DNA distribution in different cytoplasmic parasexual hybrids of *Nicotiana tabacum*. *Mol. Gen. Genet.* 165 : 231 - 237.

Belliard, G., Vedel, F., and Pelletier, G. (1979). Mitochondrial recombination in cytoplasmic hybrids of *Nicotiana tabacum* by protoplast fusion. *Nature* 281 : 401 - 403.

Binding, H., Jain, S.M., Fuiger, J., Nordhorst, G., Nehls, R., and Gressel, J. (1982). Somatic hybridisation of an atriazine resistant biotype of

Solanum nigrum with *Solanum tuberosum*. *Theor. Appl. Genet. 63* : 273 – 277.

Charbonnier, M., Faber, A.M., Pelletier, G., Belliard, G., Boutry, M., and Briquet, M. (1984). *In vitro* study of mitochondrial protein synthesis in cytoplasmic hybrids of *Nicotiana tabacum*. In *Genetic Engineering of Plants and Micro-organisms Important for Agriculture* (eds. Magnier, E. and DeNettancourt, D., Commission of European Communities, Martinus Nijhoff and Junk, Dr. W, Press, pp. 146-147.

Chen, K., Wildman, S.G., and Smith, H.H. (1977). Chloroplast DNA distribution in parasexual hybrids as shown by polypeptide composition of fraction 1 proteins. *Proc. Natl. Acad. Sci. USA 74* : 5109 – 5112.

Chetrit, P., Mathieu, C., Pelletier, G., and Primard, C. (1985). Mitochondrial DNA polymorphism induced by protoplast fusion in *Cruciferae*. *Theor. Appl. Genet. 69* : 361 – 366.

Chiu, W.L., and Sears, B. (1985). Recombination between chloroplast DNAs does not occur in sexual crosses of *Oenothera*. *Mol. Gen. Genet. 198* : 525 – 528.

Clark, E.M., Izhar, S., and Hanson, M.R. (1985). Independent segregation of the plastid genome and cytoplasmic male sterility in *Petunia* somatic hybrids. *Mol. Gen. Genet. 199* : 440 – 445.

Clayton, D.A., Teplitz, R.L., Nabholz, M., Dovey, H., and Bodmer, W. (1971). Mitochondrial DNA of human-mouse cell hybrids. *Nature, Lond. 234* : 560 – 562.

Coon, H.G., Ho, K.I., and Dawid, I. (1973). Propagation of both parental mitochondrial DNAs in rat-human and mouse-human hybrids cells. *J. Mol. Biol. 81* : 285 – 298.

Cooper-Bland, S., Kumar, A., Pirrie, A., Pental, D., and Cocking, R.C. (1986). Somatic hybridisation between *N. tabacum* (NR⁻) var. Gatersleben and *N. glutinosa* L. *Plant Cell Reports* (in press).

Cseplo, A., Nagy, F., and Maliga, P. (1984). Interspecific protoplast fusion to rescue a cytoplasmic lincomycin resistance mutation into fertile *Nicotiana plumbaginifolia* plants. *Mol. Gen. Genet. 198* : 7 – 11.

Davey, M.R., and Kumar, A. (1983). Higher plant protoplasts: Retrospect and prospect. In *Plant Protoplasts* (ed. Giles, L.K.) pp. 219 – 263 Int. Rev. Cytol. Suppl. 16 : Academic Press, New York.

De Block, M.R., Schell, J., and Montagu, M.V. (1985). Chloroplast transformation by *Agrobacterium tumefaciens*. *The EMBO Journal 4* : 1367 - 1372.

De Francesco, L. (1983) Propagation of two species of mitochondrial DNA in Chinese hamster-mouse somatic cell hybrids. *Somat. Cell Genet. 9* : 133 - 139.

De Francesco, L., Attardi, G., and Croce, C.M. (1980). Uniparental propagation of mitochondrial DNA in mouse-human cell hybrids. *Proc. Natl. Acad. Sci. USA 77* : 4079 - 4083.

Flick, C.E., and Evans, D.A. (1982). Evaluation of cytoplasmic segregation in somatic hybrids of *Nicotiana* : tentoxin sensitivity. *J. Hered 73* : 264 - 266.

Fluhr, R. (1983). The segregation of organelles and cytoplasmic traits in higher plant somatic fusion hybrids. In *Protoplasts 1983*. (eds. Potrykus, I., Harms, C.T., Hinnen, A., Hutter, R., King, P.J., and Shillito, R.D.). Lecture Proceedings, 6th Int. Protoplast Symposium, Basel, pp. 85 -92. Birkhauser Verlag, Basel.

Fluhr, R., Aviv, D., Galun, E., and Edelman, M. (1984). Generation of heteroplastidic *Nicotiana* cybrids by protoplast fusion: analysis for plastic recombination. *Theor. Appl. Genet. 67* : 491 - 497.

Fluhr, R., Aviv, D., Galun, E., and Edelman, M. (1985). Efficient induction and selection of chloroplast-encoded antibiotic resistant mutants in *Nicotiana*. *Proc. Natl. Acad. Sci. USA 82* 1485 - 1489.

Galun, E., Arzee-Gonen, P., Fluhr, R., Edelman, M., and Aviv, D. (1982). Cytoplasmic hybridisation in *Nicotiana*: Mitochondrial DNA analysis in progenies resulting from fusion between protoplasts having different organelle constitutions. *Mol. Gen. Genet. 186* 50 - 56.

Galun, E., and Aviv, D. (1983). Cytoplasmic hybridisation - genetic and breeding applications. In *Application of plant tissue culture methods for crop improvement* (D.A. Evans, W.R. Sharp, P.V. Ammirato and Y. Yamada, eds). Vol. 1., pp. 358 - 392. Macmillan Publishing Corp., New York.

Giles, K.L. (1983). Mechanisms of uptake into plant protoplasts. In *Plant cell culture in crop improvement*. (eds. Sen, S.K. and Giles, K.L.) pp. 227 - 235. Plenum, New York.

Gleba, Y.Y. (1978). Extranuclear inheritance investigated by somatic cell hybridisation. In *Frontiers of Plant Tissue Culture*. (ed. Thorpe,

T.A.) pp. 95 - 102. University of Calgary, Calgary.

Gleba, Y.Y., Butenko, R.G., Sytnik, K.M. (1975). Fusion of protoplasts and parasexual hybridisation in *Nicotiana tabacum* L. *Dokl. Akad. Nauk. USSR.* *221* : 1196 - 1198.

Gleba, Y.Y., Komarnitsky, I.K., Kolesnik, N.N., Meshkene, I., and Martyn, G.I. (1985). Transmission genetics of the somatic hybridisation process in *Nicotiana*. II. Plastome heterozygotes. *Mol. Gen. Genet.* *198* : 476 - 481.

Gleba, Y.Y., and Sytnik, K.M. (1984). *Protoplast Fusion*. Springer Verlag, Berlin.

Gressel, J., Cohen, N. and Binding, H. (1984). Somatic hybridisation of an atra resistant biotype of *Solanum nigrum* with *Solanum tuberosum*. 2. Segregation of plastomes. *Theor. Appl. Genet.* 67 : 131 - 134.

Gyllensten, U., Wharton, D., and Wilson, A. (1985). Maternal inheritance of mitochondrial DNA during backcrossing of two species of mice. *The Journal of Heredity* 76 : 321 - 324.

Hanson, M.R., and Conde, M.F. (1985). Functioning and variation of cytoplasmic genomes: Lessons from cytoplasmic-nuclear interactions affecting male fertility in plants. *Int. Rev. Cytol.* 94 : 213 - 267.

Hayashi, J., Gotoh, O., Tagashira, K., Tosu, M., and Sekiguchi, T. (1981). Analysis of mitochondrial DNA species in interspecific hybrid somatic cell using restriction endonuclease. *Exp. Cell Res. 131* : 458 - 462.

Hayashi, J., Gotoh, O., Tagashira, Y., Tosu, M., Sekiguchi, T., and Yoshida, M. (1982). Mitochondrial DNA analysis of mouse-rat hybrid cells. *Somat. Cell Genet. 8* : 67 - 81.

Hayashi, J., Takashira, Y.. and Yoshida, M.C. (1985). Absence of extensive recombination between inter- and intra-species mitochondrial DNA in mammalian cells. *Exp. Cell Res. 160* : 387 - 395.

Horak, I., Coon, H.G., and Dawid, I. (1974). Inter-specific recombination of mitochondrial DNA molecules in hybrid somatic cells. *Proc. Natl. Acad. Sci. USA 71* : 1823 - 1832.

Kemble, R.J., Brettel, R.I.S., and Flavell, R.B. (1982). Mitochondrial DNA analyses of fertile and sterile maize plants from tissue culture with the Texas male sterile cytoplasm. *Theor. Appl. Genet. 62* : 213 - 217.

Kemble, R.J., and Shepard, J.F. (1984). Cytoplasmic DNA variation in potato protoclonal population. *Theor. Appl. Genet. 69* : 211 - 216.

Kumar, A., Cocking, E.C., Bovenberg, W.A., and Kool, A.J. (1982). Restriction endonuclease analysis of chloroplast DNA in interspecies somatic hybrids of *Petunia*. *Theor. Appl. Genet. 62* : 377 - 383.

Kumar, A., and Cocking, E.C. (1986). Protoplast fusion: a novel approach to organelle genetics. *Amer. J. Bot.* (in press).

Kumar, A., Wilson, D., and Cocking, E.C. (1981). Polypeptide composition of fraction 1 protein of the somatic hybrid between *Petunia parodii* and *Petunia parviflora*. *Biochem. Genet. 19* : 255 - 262.

Lemieux, C., Tormel, M., Seligny, V.L., and Lee, R.W. (1984). Chloroplast DNA recombination in interspecific hybrids of *Chlamydomonas*: Linkage between a nonmendelian locus for streptomycin resistance and restriction fragment coding for 16S rRNA. *Proc. Natl. Acad. Sci. USA, 81* 1164 - 1168.

Levings III, C.S. (1983). The plant mitochondrial genome and its mutants. *Cell 32* : 659 - 661.

Maliga, P., Lorz, H., Lazar, C., and Nagy, F. (1982). Cytoplast-protoplast fusion for interspecific chloroplast transfer in *Nicotiana*. *Mol. Gen. Genet. 185* : 211 - 215.

Matthews, B.F., and Widholm, J.M. (1985). Organelle DNA composition and isoenzyme expression in an interspecific somatic hybrid of *Daucus*. *Mol. Gen. Genet. 198* : 371 - 376.

Medgyesy, P., Fejes, E., and Maliga, P. (1985). Interspecific chloroplast recombination in a *Nicotiana* somatic hybrid. *Proc. Natl. Acad. Sci. USA. 82* : 6960 - 6964.

Melchers, G., Sacristan, M.D., and Holder, A.A. (1978). Somatic hybrid plants of potato and tomato regenerated from fused protoplasts. *Carlsberg Res. Commun. 43* : 203 - 218.

Menczel, L., Nagy,F., Kiss, Z., Maliga, P. (1981). Streptomycin resistant and sensitive somatic hybrids of *N. tabacum* and *N. knightiana*: Correlation and resistance to *N. tabacum* plastids *Theor. Appl. Genet. 59* : 191 - 198.

Metzlaff, M., Borner, T., and Hagemann, R. (1981). Variation of chloroplast DNAs in the genus *Pelargonium* and their biparental inheritance *Theor. Appl. Genet. 60* 37 - 41.

Nagy, F., Lazar, G., Menczel, L., and Maliga, P. (1983). A heteroplasmic state induced by protoplast fusion is a necessary condition for detecting rearrangement in *Nicotiana* mitochondrial DNA. *Theor. Appl. Genet. 66* : 203 - 207.

O'Connell, M.A., and Hanson, M.R. (1985). Somatic hybridisation between *Lycopersicon esculentum* and *Lycopersicon pennellii*. *Theor. Appl. Genet. 70* : 1 - 12.

Pelletier, G., and Chupeau, Y. (1984). Plant protoplast fusion and somatic plant cell genetics. *Physiol. Veg. 22*:377 - 399.

Pelletier, G., Primard, C., Vedel, F., Chetrit, P., Remy, R., Rouselle, P., and Renard, M. (1983). Intergeneric cytoplasmic hybridisation in *Cruciferae* by protoplast fusion. *Mol. Gen. Genet. 191* : 244 - 250.

Rothenberg, M., Boeshore, M.L., Hanson,M.R., and Izhar, S. (1985). Intergenomic recombination of mitochondrial genomes in a somatic hybrid plant. *Current Genet. 9* : 616 - 619.

Sears, B. (1983). Genetics and evolution of the chloroplast. (ed. Gustafson, J.P.) In 15th Stadler Symposium. 15 : 119 - 139 University of Missouri, Columbia.

Sidorov, V.A., Menczel, L., Nagy, F., and Maliga, P. (1981). Chloroplast transfer in *Nicotiana* based on metabolic complementation between irradiated and iodoacetate treated protoplasts. *Planta. 152* : 341 - 345.

Solus, J.F., and Eisenstadt, J.M., (1984). Retention of mitochondrial DNA species in somatic cell hybrid using antibiotic selection. *Exp. Cell Res. 151* : 299 - 305.

Vasii, I.K. (1984). *Cell culture and somatic cell genetics of plants*. Academic Press, New York.

Zelcer, A., Aviv,D., and Galun, E. (1978). Interspecific transfer of cytoplasmic male sterility by fusions between protoplast of normal *Nicotiana sylvestris* and X-ray irradiated protoplasts of male sterile *N. tabacum*. *Z. Pflanzenphysiol. 90* : 397 - 407.

Zuckerman, S., Solus, J., Gillespie, F.P., Eisenstadt, J.M. (1984). Retention of both parental mitochondrial DNA species in mouse-chinese hamster somatic cell hybrids. *Somat. Cell. Genet. 10*: 85 - 92.

The fate of chloroplasts and mitochondria following protoplast fusion

13 The fate of chloroplasts and mitochondria following protoplast fusion with special reference to *Solanum* and *Brassica*

T. L. Barsby, R. J. Kemble, S. A. Yarrow and J. F. Shepard

ABSTRACT

Protoplast fusion provides a unique means to combine both nuclear and cytoplasmic genomes in a single cell. During the process of cell division and plant regeneration sorting out of these genomes takes place so that normally the somatic hybrid may not contain the total genomic complement of both parents. The ability to characterize chloroplasts and mitochondria by restriction enzyme analysis enables the ultimate fate of these organelles in the regenerated plant to be deterimined. The analysis of fusion products between somatic hybrids of *Solanum tuberosum* and *S. brevidens* illustrates some of the possible outcomes. These analyses were extended to fusion products within *Brassica* and have provided several novel sources of *B. napus* germplasm.

INTRODUCTION

Protoplast fusion presents a method of combining diverse nuclear and cyto-plasmic genomes within the same cell. At some time between initial division and plant regeneration, predominance of a particular cytoplasmic genome occurs. It is assumed that in most cases organelle sorting out is random except when selection pressure for a particular organelle type has applied. Recent studies have revealed that organelles, or at least mitochondria, do not always remain unaltered during this process (Hanson, 1984).

Creation of new nuclear-cytoplasmic combinations has been limited, (like many tissue-culture dependent phenomena), by the ability to recover whole plants from fusion products. Only in those situations where plant regeneration is possible, and at a workable frequency, can generalisation be made (see Pelletier, this volume). It is not surprising, therefore, that most of the observations and generalisations made to date have been with *Nicotiana* species (e.g., Belliard, Vedel and Pelletier, 1979); Nagy

Organelle genomes in fusion products

Torok and Maliga, 1981, 1983; Galun, Arzee-Gonen, Fluhr, Edelman and Aviv, 1982; Aviv, Arzee-Gonen, Bliechman and Galun, 1984), where protoplast culture and regeneration are well controlled.

There are only recently published reports of somatic hybrids or cybrids with ct-DNA which differs from that of either parent or of a mixture of both parents (see Pelletier, this volume). Mitochondrial DNA, however, has been shown to undergo recombination in several somatic hybrids and cybrids (e.g. Belliard et al., 1979; Nagy et al., 1981; Boeshore, Lifshitz, Hanson and Izhar, 1984; Pelletier, Primard, Vedel, Chetrit, Remy, Rouselle and Renard, 1983).

CHLOROPLAST AND MITOCHONDRIAL FATES IN SOMATIC FUSION PRODUCTS

The investigations to be described here involve somatic protoplast fusions involving *Solanum* and *Brassica* species. The experiments were based upon the regeneration system developed for *S. tuberosum* (Shepard and Totten, 1977; Shepard et al., 1980) and later shown to be applicable, with some modification, to *S. brevidens* (Barsby and Shepard, 1983) and to *B. napus* (Barsby et al., in preparation). All of the fusion products recovered to plants exhibited unaltered ct-DNA patterns. In some cases the mitochondria were deemed to have undergone several types of genomic rearrangements.

S. tuberosum and *L. esculentum*: Intergeneric somatic hybrids
The first fusion experiments in which intergeneric somatic hybrids between potato and tomato were developed were undertaken by Melchers, Sacristan and Holder, 1978. Mesophyll protoplasts of a diploid yellow-green mutant of *L. esculentum* were fused with protoplasts derived from a liquid callus culture of the dihaploid strain "HH258" of *S. tuberosum*. Hybrids were selected on the non-regenerability of the tomato line and on morphological differences between hybrids and the potato parent. Chloroplast DNA analysis revealed no mixtures of chloroplast types. The resulting hybrids were fused 'pomatoes' (hybrids with tomato chloroplasts) and 'topatoes' (hybrids with potato chloroplasts).

Similar hybrids were reported by Shepard, Bidney, Barsby and Kemble (1983), but different tissue sources and culture methods were used. Mesophyll protoplasts of a yellow potato variant derived from a protoclone of tetraploid (4n=48) *S. tuberosum* cv. 'Russet Burbank' (Shepard, 1980)

were fused with mesophyll protoplasts of diploid *L. esculentum* cvs. 'Rutgers' or 'Nova'. Hybrids were selected on the basis of cultural differences between the parents. The four hybrids which were produced had only potato chloroplasts and mitochondria, the additive chromosome number (72) and contained fraction 1 protein profiles expected in true amphidiploid somatic hybrids.

S. tuberosum and *S. brevidens*: Interspecific hybridisations

Somatic hybridisation between more distantly-related species resulted in the production of hybrids between tetraploid (4n=48) *S. tuberosum* (the cultivated potato) and diploid (2n=24) *S. brevidens* (a non-tuberous wild relative). The primary purpose of these experiments was to determine whether it was possible to produce hybrids which were not directly attainable by conventional crossing, and which could provide a source of resistance to certain viruses and to frost. The plants in question have not been tested for the presence or absence of these characters; but recently

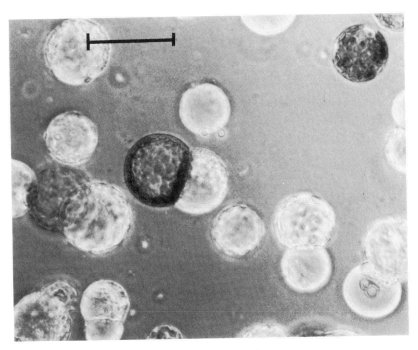

Fig. 1 - Fusion of *S. brevidens* mesophyll protoplasts and albino
         S. tuberosum protoplasts in the presence of poly-
         ethylene glycol. (Bar=100 m).

similar types of hybrids have been shown to carry resistance to potato leaf roll virus (Austin, Baer and Helgeson, 1985).

Protoplasts were isolated from mesophyll tissue of wild-type *S. brevidens* and from an axenically maintained albino protoclone of *S. tuberosum* cv. 'Russet Burbank'. Fig. 1 shows the first step in the hybridisation process. The production of somatic hybrids has been described previously (Barsby, Shepard, Kemble and Wong, 1984). Hybrids were selected on the basis of their green colour and the ability to regenerate plants on a medium which was non-inductive for shoots of *S. brevidens*. Twenty-five calli formed shoots (Fig. 2), two forming two plants, and one forming three plants, giving a total of twenty-nine plants (Fig. 3).

Fig. 2 - Multiple shoot regeneration from somatic fusion-derived callus

Fig. 3 - A *S. tuberosum-S. brevidens* somatic hybrid.

The plants were analysed for their chromosome complements, the small subunit of ribulose bisphosphate carboxylase-oxygenase (Fraction 1 protein) and chloroplast and mitochondrial DNA (see the methods used for ct-and mt-DNA preparation and analysis, Kemble and Shepard (1984). All of the plants analysed had the Fraction 1 profiles of both parents. Chromosome numbers varied between the additive (72) and the octoploid condition (96). The chloroplast genome of *S. brevidens* was evident in all the hybrids produced in this experiment. Green plants have never been recovered from protoplasts of the albino potato parent, so it was expected that chlorophyll containing hybrids would possess *S. brevidens* ct-DNA. Indeed no variation from the parental *S. brevidens* ct-DNA profile was detected (Fig. 4).

One of the regenerants, even though albino was otherwise morphologically similar to chlorophyll-containing hybrid regenerants. It is possible that this plant resulted from the regeneration of an unfused *S. tuberosum* protoplast. However, this was unlikely because of the type of selection process employed, and especially since protoplasts of this material have not been shown to regenerate plants under the conditions which are employed in

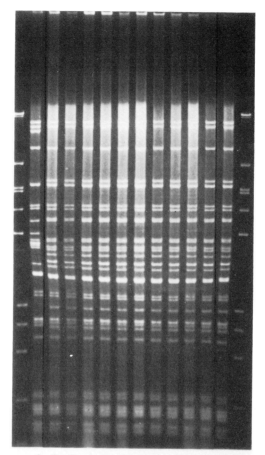

m 1 2 3 4 5 6 7 8 9 10 11 12 m

Fig. 4 - Electrophoresis on 1% agarose gel of *Bam* HI fragmented
        ct-DNA from: lane 1, Parental *S. tuberosum* protoclone;
        lane 2, parental *S. brevidens*; lanes 3-12, ten different
        S. tuberosum - *S. brevidens* somatic hybrids. Lane 'm'
        contains size marker fragments produced by independent
        lambda DNA with *Eco* RI and *Hae* III.

this experiment. Unfortunately since small amounts of leaf material were
available, ct-DNA of this plantlet was not analysed. Therefore the
possibility that this plantlet could have retained the chloroplast of the
potato parent, thus producing an albino phenotype, could not be proved
un-equivocally.

A similar type of experiment was performed in which the *S. brevidens* protoplasts were treated with iodoacetic acid prior to fusion, and the *S. tuberosum* was the yellow variant used in the potato-tomato fusions described above (rather than the albino mutant). The single plant recovered had additive Fraction 1 profiles, and 72 chromosomes, but the ct-DNA of *S. tuberosum*. The difference in chloroplast type (as compared to hybrids from the first experiment) may be attributed to the action of the iodoacetic acid on the cytoplasm or to the introduction of the more compatible plastid of the potato parent. When protoplasts of this yellow variant are cultured, green revertant calli sometimes occur and will readily regenerate plants, whereas non-green calli do not regenerate under the same culture conditions.

The mt-DNA of the hybrids was analysed with seven different restriction enzymes because use of only one enzyme may not necessarily reveal any changes in the profile, e.g. the *Hind* III digestion produces (Fig. 5). Three hybrids regenerated from the same callus had mt-DNA profiles identical to those of unaltered *S. brevidens,* whereas digestion with *Bam* HI (Fig. 6), clearly showed differences in mt-DNA arrangement.

The mt-DNA of the hybrids was more diverse in origin and nature than that of the chloroplasts. Some hybrids possessed the unchanged mitochondrial genome of *S. tuberosum*, others showed evidence of intra-mitochondrial recombination and some had slight alterations in either parental genome.

None of the hybrids displayed mt-DNA restriction patterns characteristic of a mixture of both parental types, which is further evidence that the plants were true hybrids and not chimeras. Twenty-nine plants were analysed including some from a common callus origin. Fourteen hybrids retained either unaltered or slightly altered potato mitochondrial genome organization. The other fifteen possessed a non-parental mitochondrial genome.

Mitochondrial recombination is indicated in restriction patterns by the presence of some bands of each parent, the absence of particular bands of each parent, and the existence of new non-parental bands. Twelve hybrids possessed patterns which indicated that mitochondrial recombination phenomenon had occurred. In addition, three of the hybrid plants produced similar patterns, without any new, non-parental bands, i.e. these somatic hybrids contained a mixture of both parental genomes with respective

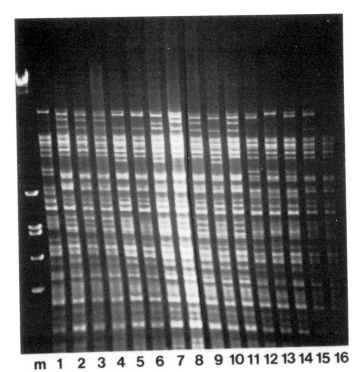

Fig. 5 - Electrophoresis on 0.7% agarose gel of *Hind* III fragmented
mt-DNA from: lane 1, parental *S. brevidens;* lane 2,
S. tuberosum protoclone; lanes 3-15, thirteen different
S. tuberosum-S. brevidens somatic hybrids. Lane 'm'
contains size marker fragments of lambda DNA fragmented
with *Eco*RI. DNA fragments smaller than approximately
3.5 kb are not shown.

deletions in both. These results demonstrated that recombinant activity of
some sort must have occurred early or as a result of the protoplast fusion
process. It is suggested that the currently accepted explanations for
mitochondrial recombination are incomplete and that other events super-
imposed on initial recombination events could possibly result in the altered
patterns. In the above case recombination occurs between parental types
(either as a result of the primary recombination event or after several cycles
of genome replication) which is followed by a deletion of portions of each

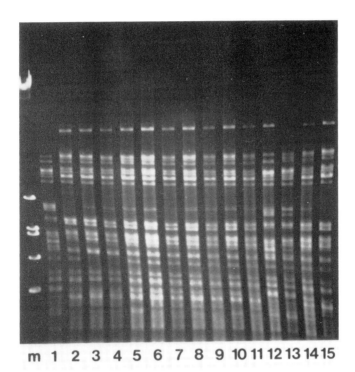

m 1 2 3 4 5 6 7 8 9 10 11 12 13 14 15

Fig. 6 - Electrophoresis on 0.7% agarose gel of *Bam* HI fragmented
mt-DNA from: lane 1, parental *S. brevidens*; lane 2,
parental *S. tuberosum* protoclone; lanes 3-16 fourteen
different *S. tuberosum-S. brevidens* somatic hybrids.
Lane 'm' size marker fragments of lambda DNA fragmented
with *Eco*RI. DNA fragments smaller than approximately
3.5 kb are not shown.

mitochondrial genome. It is possible that there may be a mitochondrial
genome size restraint which is species specific. Alternatively, each
parental genome may have recombined within itself *prior to* mitochondrial
fusion, resulting in a deleted genome. Evidence for this type of process
occurring in potato protoclones has already been presented by Kemble and
Shepard (1984). Examination of mitochondrial genomes of forty-seven
protoclones of *S. tuberosum* cv. 'Russet Burbank' revealed that as many as
fifteen percent showed mitochondrial rearrangements. This type of tissue-
culture induced recombination in potato plants from *unfused* protoplasts

might also explain some of the restriction patterns observed in the somatic hybrids. Another explanation which is strongly suggested by the restriction patterns shown in Fig. 3 is that the modified mitochondrial genomes co-exist in the cytoplasm of the hybrids, a 'mitochondrial chimera' of numerically equal populations.

Many of the new non-parental bands referred to above were common to several hybrids, indicating that recombination was not random and that recombination 'hot-spots' most likely exist on the mitochondrial chromosome. This theory is expanded by Vedel (this volume). Further understanding of these regions may be enhanced by the cloning of regions of the parental mitochondrial genomes and the use of these clones to probe the complete mitochondrial genome or restricted chromosome fragments.

One further important result of this analysis is that different plants regenerated from the *same* callus need not necessarily possess the same mt-DNA restriction profiles. One individual hybrid possessed normal potato mitochondria while its 'twin' had a recombinant type. In another case, a deleted potato pattern occurred in one plant, and a normal potato pattern in its twin. One possible explanation for phenomena of this type is that the two plants arose from two different hybrid protoplasts which developed in juxtaposition. This is considered to be unlikely as firstly, it is a relatively small proportion of fusion products that eventually develop in the post-fusion population, so that the chances of such adjacent development occurring are extremely low and secondly, the culture procedures used would tend to select against clusters of different cell lineages. The regeneration of plants with different mitochondria from the same callus may be an indication that predominance of one mitochondrial type does not necessarily occur until shoots are differentiated from the callus.

Hybrids such as those described above are of limited value in plant breeding, and recently much attention has been directed toward the cytoplasmic recombinations made possible through protoplast fusion (see Pelletier, this volume). There has been considerable interest in the genus *Brassica*. The interest in our group is in the combination of male sterility and triazine tolerance in the same cytoplasm by protoplast fusion. Such cybrids have been successfully produced by Pelletier *et al.*, (1983).

B. napus and *B. campestris*: Interspecific Hybridisation

In order to produce somatic hybrids and cybrids, there are several essential requirements which must be met. These are (1) plant regeneration from protoplasts, (2) protoplast fusion and (3) a suitable means of hybrid selection and screening and an appropriate definitive method of hybrid characterisation. Although there are many published reports of plant regeneration from protoplasts of a particular species, it is no small matter to establish this technology in a consistent manner.

The first step in cybrid recovery is the development of a suitable system for protoplast culture and plant regeneration. This has been accomplished in our laboratory by applying the approaches used on the potato systems described above. Protoplasts were isolated from hypocotyls of *B. napus* cv. 'Regent', (Fig. 7), carrying 'polima' male sterile (mitochondria-encoded) cytoplasm and from leaf mesophyll of *B. campestris* cv. 'Candle' carrying triazine-tolerant (chloroplast-encoded) cytoplasm.

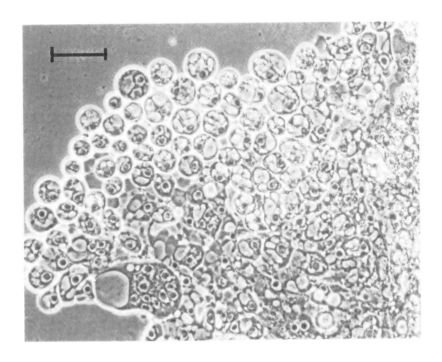

Fig. 7 - Liberation of protoplasts from hypocotyls of *Brassica napus*. (Bar=100 μm).

Organelle genomes in fusion products

Fusion was effected as described previously (Barsby and Shepard 1983).
B. campestris protoplasts will not regenerate plants in the system used
for *B. napus*. *B. napus* protoplasts were treated with 2 mM iodoacetic acid
prior to fusion to prevent division. Consequently, regenerants were
likely to be from a non-parental source. It is expected that some of the
regenerants will contain the desired combination of chloroplasts and
mitochondria since these organelles should assort randomly following
protoplast fusion. In this particular parental combination it is also
likely that hybrid nuclei will result in the progeny.

    Cybrid plants were recovered, some of which had 'polima' male sterile
mitochondria and 'triazine tolerant' chloroplasts. This demonstrated that
this cytoplasmic recombination was possible by protoplast fusion. Cybrids
recovered using this parental combination had not been fully analysed at
the time of writing but there are indications of low female fertility
possibly caused by the presence of a hybrid nucleus.

B. napus and *B. napus*: Intraspecific hybridisation

The experiments described above have now been extended to utilise *B. napus*
as parents into which the triazine-tolerant cytoplasm has been introduced
by repeated backcrossing. Protoplasts of this line are gamma-irradiated
prior to fusion. These two measures reduce the possibility of hybrids or
polyploids occurring in the progeny. Cybrids have been recovered with the
desired organelles, and it is expected that these will prove to be female
fertile.

FUTURE PROSPECTS

These latter experiments have been primarily designed to produce agronomically important *B. napus* lines. In addition, methodology is now being
established which is simple, efficient and reproducible. Experiments
comparing various screening and selection techniques should also enable the
direction of fusion towards a particular mitochondrial or chloroplast type.
An important additional benefit might be to enable the production of cybrids
with more or less stable mitochondrial genomes. Tissue culture *per se*
can affect mitochondrial DNA restriction enzyme profiles (Kemble and
Shepard, 1984), a fact which stresses the importance of appropriate analyses

of control material (i.e. the protoclones derived from parental lines).
It is currently concluded tentatively that the *B. napus* mitochondrial
genome is relatively stable as no mt-DNA changes have been detected during
the analysis of numerous plants recovered from unfused protoplasts. In
addition, no evidence of mitochondrial recombination has yet been found
in any of the cybrid progeny.

CONCLUSIONS

The random association of different chloroplast and mitochondrial populations
following fusion suggests opportunities for recombining cytoplasmic traits,
thus expanding the range of cytoplasmic diversity. This also allows the
manipulation of organellar traits, such as male sterility, to agronomic
advantage. The creation of novel nuclear-cytoplasmic combinations provides
an opportunity to contribute to the understanding of organelle genome
structure, organisation and function. There is no evidence yet that tissue
culture-induced effects on mitochondria influence plant phenotype. However,
as greater understanding of the mitochondrial genome is gained, change may
be viewed either as an interesting phenomenon with 'potential', or as a
problem in the application of this technology. The ideal situation is one
in which there is control of stability or instability in genomes. It is
interesting to note that the differences in observed degree of genome
alterations following protoplast fusion and regeneration reported by
different workers may in fact be due to differences in the techniques used
by different research groups to achieve fusion (i.e. whether by chemical
or electrical means) or to achieve regeneration (i.e. the types of osmotica
and growth regulations used in culture media).

REFERENCES

Austin, S., Baer, M.A. and Helgeson, J.P. (1985). Transfer of resistance to
    potato leaf roll virus from *S. brevidens* into *S. tuberosum* by somatic
    fusion. *Plant Science* 39, 75-82.
Aviv, D., Arzee-Gonen, P. Bliechman, S., and Galun, E. (1984). Novel
    alloplasmic *Nicotiana* plants by "donor-recipient" protoplast fusions:
    cybrids having *N. tabacum* or *N. sylvestris* nuclear genomes and either

or both plastomes and chondriomes from both species. *Mol. Gen. Genet.*
196, 244-253.

Barsby, T.L., and Shepard, J.F. (1983). Regeneration of plants from
protoplasts of *Solanum* species of the *Etuberosa* group. *Plant Sci. Lett.*
31, 101-105.

Barsby, T.L., Yarrow, S.A., and Shepard, J.F. (1984). Heterokaryon
identification through simultaneous fluorescence of tetramethylrhodamine
isothiocyanate and fluorescein isothiocyanate labelled protoplasts.
Stain Tech. 59, 217-220.

Barsby, T.L., Shepard, J.F., Kemble, R.J., and Wong, R. (1984). Somatic
hybridisation in the genus *Solanum: S. tuberosum* and *S. brevidens*
Plant Cell Reports 3, 165-167.

Belliard, G., Pelletier, G., Vedel, F., and Quetier, F. (1978). Morphological characteristics and chloroplast DNA distribution in different
cytoplastic parasexual hybrids of *N. tabacum. Mol. Gen. Genet. 165,*
231-237.

Belliard, G., Vedel,F., and Pelletier, G. (1979). Mitochondrial recombination in cytoplasmic hybrids of *N. tabacum* by protoplast fusion. *Nature*
281, 401-403.

Boeshore, M.L., Lifshitz, I., Hanson, M.R., and Izhar, S. (1984). Novel
composition of mitochondrial genomes in *Petunia* somatic hybrids derived
from cytoplasmic male sterile and fertile plants. *Mol. Gen. Genet. 190,*
459-467.

Galun, E., Arzee-Gonen, P., Fluhr, R., Edelman, M., and Aviv, D. (1982).
Cytoplasmic hybridisation in *Nicotiana:* mitochondrial DNA analysis in
progenies resulting from fusion between protoplasts having different
organelle constitutions. *Mol. Gen. Genet. 198*, 476-481.

Hanson, M. (1984). Stability, variation and recombination in plant
mitochondrial genomes via tissue culture and somatic hybridisation.
In *Oxford Surveys of Plant Molecular and Cell Biology* (ed. Miflin, B.J.)
pp 33-52, Clarendon Press, Oxford.

Kemble, R.J., and Shepard, J.F. (1984). Cytoplasmic DNA variation in a
potato protoclonal population. *Theor. Appl. Genet 69*, 211-216.

Melchers, G., Sacristan, M., and Holder, A. (1978). Somatic hybrid plants
of potato and tomato regenerated from fused protoplasts. *Carlsberg Res.*
Commun. 43, 203-218.

Nagy, F., Torok, I. and Maliga, P. (1981). Extensive rearrangements in the mitochondrial DNA in somatic hybrids of *N. tabacum* and *N. knightiana*. *Mol. Gen. Genet. 183,* 437-439.

Nagy, F., Lazar, G., Menczel, L., and Maliga, P. (1983). A heteroplasmic state induced by protoplast fusion is a necessary condition for detecting rearrangements in *Nicotiana* mitochondrial DNA. *Theor. Appl. Genet. 66,* 203-207.

Pelletier, G., Primard, C., Vedel, F., Chetrit, P., Remy, R., Rouselle, P., and Renard, M. (1983). Intergeneric cytoplasmic hybridisation in *Cruciferae* by protoplast fusion. *Mol. Gen. Genet. 191,* 244-250.

Shepard, J.F. (1980). Mutant selection and plant regeneration from potato mesophyll protoplasts. In *Genetic Improvement of Crops: Emergent Techniques*. (eds. Rubenstein, I., Gengenbach, B., Phillips, R., and Green, E.) pp 185-219, University of Minnesota Press, Minneapolis.

Shepard, J.F., and Totten, R.E. (1977). Mesophyll cell protoplasts of potato. *Plant Physiol. 60,* 313-316.

Shepard, J.F., Bidney, D., and Shahin, E. (1980). Potato protoplasts in crop improvement. *Science 208,* 17-24.

Shepard, J.F., Bidney, D., Barsby, T.L., and Kemble, R.J. (1983). Genetic transfer in plants through interspecific protoplast fusion. *Science 219,* 683-688.

14   Chloroplast genomes in hybrid calli derived from cell fusion: a novel system to study chloroplast segregation in hybrid calli

S. Akada and A. Hirai

ABSTRACT

Nicotiana glauca possessing *N. gossei* chloroplasts was produced by sexual hybridisation and a back cross. Protoplasts from this plant were fused with *N. langsdorffii* protoplasts to study the mode of chloroplast separation in parasexual hybrid cells. This combination of cell fusion is convenient for selecting hybrids at the stage of small calli by expression of genetic tumors, and distinguishing two kinds of chloroplast easily by the analysis of Fraction I protein and restriction enzyme patterns of chloroplast DNA. This paper shows that it is a good system with which to study the mode of chloroplast separation in parasexual hybrid cells.

INTRODUCTION

Cell fusion between two different types of cells brings about the mixture of two kinds of chloroplasts as well as the fusion of two nuclei in a new cell. Thus, it may be possible to produce new plants which contain two kinds of chloroplasts or hybrid chloroplasts. However, although some workers have reported the evidence of parasexual hybrid plants containing two kinds of chloroplasts (Gleba, 1978; Glimelius, Chen and Bonnet, 1981). Only one kind of chloroplast was detected from hybrid plants in many cases (Chen, Wildman and Smith, 1977; Melchers, Sacristan, and Holder, 1978; Douglas, Wetter, Keller, and Setterfield, 1981).

To obtain information on how to explain whereby the two chloroplast populations separate in the process of formation of hybrid plants from fused cells, it is important to know how two kinds of chloroplasts behave during the growth from a fused cell to the callus which is ready to differentiate shoots.

Therefore, protoplasts from *Nicotiana glauca* and *N. langsdorffii* were fused and hybrid calli selected in a hormone-free medium. Sexual hybrids of these two species spontaneously form tumours which are able to grow in tissue culture medium without any plant hormones. This phenomenon allows a selective recovery of parasexual hybrid cells from parental cells (Shaeffer and Smith, 1963, Carlson, Smith, and Learing, 1972). Then the chloroplast genomes in hybrid calli were analysed using the large sub-unit of Fraction I protein (ribulose 1.5-bisphosphate carboxylase) as a chloroplast genetic marker. The results showed that 70% of unseparated original hybrid calli contained two kinds of chloroplast genomes (Akada, Hirai and Uchimiya, 1983). This is contrary to previously reported results of analyses of chloroplast genome behaviour present in leaves from differentiated parasexual hybrid plants of the same species.

However, when we examined calli in detail, the ratio of the two kinds of chloroplasts were completely different in the position of each hybrid callus. For example, one part of a callus contained *N. langsdorffii* chloroplasts in 1 : 9 ratio but the other part of the same callus contained them in 9 : 1 ratio (Hirai, 1982). We have also analysed chloroplast genomes in single cells from which originated the hybrid callus, and showed that about 80% of cells contained only one or the other kind of chloroplasts but another 20% of cells contained two parental kinds of chloroplasts in a cell. These results suggested that two kinds of chloroplasts are randomly separated by cell fusions at least in the state of callus, but more information is needed to confirm this hypothesis.

The reliability of Fraction I protein as a chloroplast genetic marker was also checked using restriction endonuclease pattern of ct-DNA as the second chloroplast marker (Ichikawa, Akada and Hirai, 1984). Because it is not certain that the callus in which only one kind of Fraction I protein was expressed has only that one kind of chloroplasts, it is possible that the callus still had two kinds of chloroplasts. However, the gene for only one kind of the large subunit of Fraction I protein was expressed. There-fore, chloroplast genomes in 16 hybrid calli were analysed by both Fraction I proteins and *Bam* HI patterns of ct-DNA. The results obtained using these two markers were in agreement without exception. It showed that Fraction I protein is a good marker for this kind of work, since it needs only a small amount of tissue and is easy to analyse.

However, the pI's of the large subunit of Fraction I protein from
N. glauca and *N. langsdorffii* are only different 0.1 unit and it is
difficult to distinguish them by isoelectrofocussing especially when there
are two kinds of large subunits present in different amounts. Australian
species of *Nicotiana* have the most basic large subunit of Fraction I
protein in the genus. Therefore, it is better to fuse Australian species of
Nicotiana such as *N. gossei* and *N. langsdorffii* and to compare two kinds
of chloroplasts when Fraction I protein is used as a marker. On the other
hand, the use of *N. glauca* and *N. langsdorffii* for the parents of cell
fusions provides the opportunity by which to select hybrid calli when
calli are still in small pieces in the hormone-free medium.

The production of a plant which has *N. gossei* chloroplasts and whose
nucleus was mostly from *N. glauca* was attempted. Protoplasts from this
plant and *N. langsdorffii* protoplasts were fused. This system is now
described since it is useful for obtaining more information about the mode
of separation of chloroplast genomes in hybrid calli.

METHODS AND MATERIALS

Plant materials

Plant growing conditions and the method used for propagation of cultured
cells were reported previously (Akada *et al.*, (1983).

Cell fusion

Protoplasts from different species were fused by polyethylene glycol method
(as used in the above study).

Analysis of Fraction I protein and chloroplast DNA

Fraction I protein (Hirai, 1982) and ct-DNA (Ikikawa, *et al.*, 1984) were
analysed by the method previously reported in Akada *et al.*, (1983).
Southern transfer was conducted by the method of Shinozaki and Sugiura
(1982).

RESULTS AND DISCUSSIONS

Production of *N. glauca* containing *N. gossei* chloroplasts

Freshly matured pollen in detached *N. glauca* flowers was placed in physical contact with the stigma of attached *N. gossei* flowers whose anthers had been previously emasculated. Seeds obtained were grown in a green house, and checks made to confirm whether the plants were true hybrid by Fraction I protein analysis. Protoplasts from the true hybrids were fused with the protoplasts from *N. langsdorffii* and attempts were made to select para-sexual hybrid calli in hormone-free medium. However, no calli grew on this medium.

   N. glauca pollen was again placed in physical contact with the stigma of the hybrid plant and a B_1F_1 plant was obtained. A flow diagram showing the procedure used to produce this plant is shown in Fig. 1. Protoplasts from the back crossed progeny plant and from *N. langsdorffii* were fused and cultured in hormone-free medium to select parasexual hybrid calli. Many calli were selected which grew in hormone free medium, and Fraction I pro-tein analysis showed that these calli were hybrid between the back-crossed plant and *N. langsdorffii*. This result shows that if a little more than half of the nucleus comes from *N. glauca,* the plant can express the genetic tumour in cooperation with the *N. langsdorffii* nucleus, whereas presence of less than half of the *N. glauca* nucleus can not. This finding is interesting when related with Ahuja's hypothesis (Ahuja, 1968), that one should be able to obtain full tumour expression on the hybrid derivatives carrying one or few *N. langsdorffii* chromosomes against a diploid *N. glauca* background.

Nature of parasexual hybrid calli between the back-crossed plant and *N. langsdorffii*

Parasexual hybrid cells between the back-crossed plant and *N. langsdorffii* grew vigorously in the medium which did not contain any phytohormones, and developed into green calli which were indistinguishable from *N. glauca-N. langsdorffii* parasexual hybrid calli. As shown in Fig. 2, Fraction I protein was analysed in each callus, and they were proven that they were real hybrid calli by the small subunit of the protein. The large subunits of *N. gossei* and *N. langsdorffii* are distinctive, and the results show that in many cases, they contained both *N. gossei* and *N. langsdorffii* chloroplast genomes. However, the ratio of two kinds of chloroplasts estimated by Fraction I protein were very different in the relative positions of a callus.

Fig. 1 – Production of *N. glauca* with *N. gossei* chloroplasts
        by sexual hybridisation and a backcross.

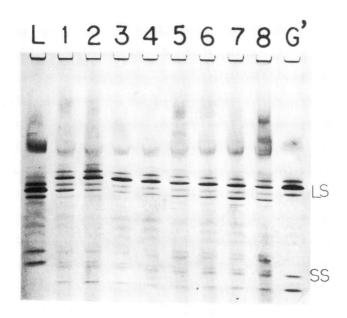

Fig. 2 - Isoelectrofocussing analysis of Fraction I proteins
obtained from *N. langsdorffii* (L). The backcrossed
plant (G) and parasexual hybrid calli (1-8). LS. large
subunit polypeptides: SS. small subunit polypeptides.

This evidence also confirmed the previous results obtained from *N. glauca-
N. langsdorffii* parasexual hybrid calli (Akada and Hirai, 1985).

Chloroplast DNAs were isolated from *N. gossei* and *N. langsdorffii* and
digested with restriction enzyme *Bam* HI and the fragments were fractionated
by agarose gel electrophoresis. As shown in Fig. 3, the size of the third
fragment was different in two species. Chloroplast DNA from *N. gossei*
has a extra *Bam* HI site in *Bam*-3 fragment of *N. tabacum* and *N. langsdorffii*.
This is also shown in Fig. 3 by the result of radioautography using the
Bam 3 fragment of *N. tabacum* (a gift from Dr. Sugiura, Nagoya University) as
probe. According to Tassopulu and Kung (1984), the third fragment of *Bam*
HI digestion locates about 15 kbp away from the gene of the large subunit
of Fraction I protein. Therefore the third fragment could be used as the
second marker for comparison between *N. gossei* and *N. langsdorffii* chloroplasts.

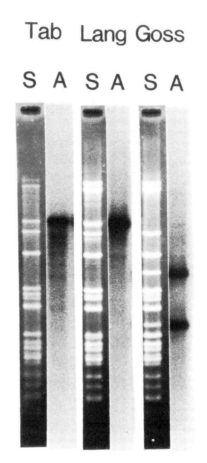

Fig. 3 - Hybridisation of *N. tabacum Bam*-3 fragment to *Bam* HI
        digested fragments of chloroplast DNA from *N. tabacum*
        (Tab). *N. langsdorffii* (Lang) and *N. gossei* (Goss).
        S:Stained by EtBr; A:Autoradiograph.

Chloroplast genomes in this hybrid calli were analysed by both Fraction
I protein and ct-DNA, and the results showed that the kinds of chloroplast
genomes indicated by use of the two markers coincided . It indicated that
recombination of ct-DNA did not occur by cell fusion in these experiments,
and also agreed with the results of previous work on the *N. glauca* -
N. langsdorffi parasexual hybrid.

    These results clearly showed that the B_1F_1 hybrid *N. langsdorffii*
parasexual hybrid is similar to the *N. glauca-N. langsdorffii* hybrid.

Moreover, the former has two kinds of chloroplasts which we can distinguish easily by both Fraction I protein and ct-DNA restriction patterns. Since this hybrid callus also provides the **same** opportunity by which hybrid cells can be selected in hormone-free medium, this particular combination is an ideal system with which to study the mode of chloroplast genome development in parasexual hybrid plants.

REFERENCES

Ahuja, M.R. (1968). An hypothesis and evidence concerning the genetic components controlling tumour formation in *Nicotiana*. *Molec. Gen. Genet. 103*, 176-184.

Akada, S., and Hirai, A. (1983). Studies on the mode of separation of chloroplast genomes in parasexual hybrid calli. II. Heterogeneous distribution of two kinds of chloroplast genomes in hybrid callus. *Plant Sci. Lett. 32*, 95-100.

Akada, S., Hirai, A., and Uchimiya, H. (1983). Studies on the mode of separation of chloroplast genomes in parasexual hybrid calli. I. Fraction I protein composition in unseparated hybrid callus. *Plant Sci. Lett. 31*, 223-230.

Carlson, P.S., Smith, H.H., and Dearing, R.D. (1972). Parasexual interspecific plant hybridisation. *Proc. Natl. Acad. Sci. USA 69*, 2292-2294.

Chen, K., Wildman, S.G., and Smith, H.H. (1977). Chloroplast DNA distribution in parasexual hybrids as shown by polypeptide composition of Fraction I protein. *Proc. Natl. Acad. Sci. USA 74*, 5109-5112.

Douglas, G.C., Wetter, L.R., Keller, W.A., and Setterfield, G. (1981). Somatic hybridisation between *Nicotiana rustica* and *N. tabacum*. IV. Analysis of nuclear and chloroplast genome expression in somatic hybrids. *Can. J. Bot. 59*, 1509-1513.

Gleba, Y.Y. (1978). Extranuclear inheritance investigated by somatic hybridisation. In *Frontiers of Plant Tissue Culture* 1978. 95-102.

Glimelius, K., Chen, K., and Bonnet, H.T. (1981). Somatic hybridisation in *Nicotiana*: Segregation of organellar traits among hybrid cybrid plants. *Planta 153*, 504-510.

Hirai, A. (1982). Isoelectrofocusing of non-carboxymethylated Fraction I protein from green callus. *Plant Sci. Lett. 25*, 37-41.

Ichikawa, H., Akada, S., and Hirai, A. (1984). Correlation between the type species of chloroplast DNA and that of the large subunit of Fraction I protein in parasexual hybrid calli. *Jpn. J. Genet. 59,* 315-322.

Melchers, G., Sacristan, M.D., and Holder, A.A. (1978). Somatic hybrid plants of potato and tomato regenerated from fused protoplasts. *Carlsberg Res. Commun. 43,* 203-218.

Schaeffer, G.W., and Smith, H.H. (1963). Auxin-kinetin interaction in tissue cultures of *Nicotiana* species and tumour conditioned hybrids. *Plant Physiol. 38,* 291-297.

Shinozaki, K., and Sugiura, M. (1982). The nucleotide sequence of the tobacco chloroplast gene for the large subunit of ribulose-1,5-bisphosphate carboxylase/oxygenase. *Gene 20,* 91-102.

Tassopulu, D., and Kung, S.D. (1984). *Nicotiana* chloroplast genome. Deletion and hot spot - a proposed origin of the inverted repeats. *Theor. Appl. Genet. 67,* 185-193.

15 The synthesis of gametosomatic hybrids between *Nicotiana tabacum* and *N. glutinosa* by protoplast fusion

A. Pirrie, J. B. Power and E. C. Cocking

ABSTRACT

Novel 'gametosomatic' hybrid plants were obtained following fusions between nitrate reductase deficient (NR⁻) *Nicotiana tabacum* leaf mesophyll protoplasts (2n) and *N. glutinosa* tetrad protoplasts (n). Hybridity was confirmed morphologically, and from esterase and peroxidase zymograms. Hybrids were found to possess the expected pentaploid (functionally triploid) chromosome number of $3n = 5x = 60$.

Preliminary results of Fraction I protein analysis indicate that hybrids possess the large subunit (chloroplast encoded) of *N. tabacum* and *N. glutinosa*. Progeny obtained in crosses with *N. tabacum* using the 'gametosomatic' hybrid as female partner were found to be tentoxin resistant, characteristic of the *N. tabacum* chloroplast type.

INTRODUCTION

Early in pollen development meiosis occurs in the pollen mother cells giving rise to tetrads consisting of four haploid spores bound within a thick callose wall (Bennett 1976). Tetrad protoplasts can be isolated by the enzymatic dissolution of the callose wall (Bohjwani and Cocking, 1972), but few studies have been performed using tetrad protoplasts, largely due to the observed lack of sustained division in cultured tetrad protoplasts.

Recently there has been renewed interest in the use of haploid protoplasts, following the proposal that the synthesis of triploid somatic hybrid plants might facilitate limited gene transfer (Pental and Cocking, 1985). At meiosis the haploid chromosome set of an allotriploid may be lost, or the chromosomes randomly segregate, such that by repeated back crossing to the diploid fusion partner the complete haploid chromosome set would be eliminated. If introgression of any traits from the haploid chromosome set into the diploid chromosome set had occurred, then this may be detected in the subsequent progeny.

Haploid protoplasts can be isolated from anther culture-derived haploid plants or tissues (Negrutiu, 1980). However, the range of species in which anther culture is successful is limited (Maheshwari, Rashid and Tyagi, 1982). Tetrad protoplasts have been isolated from a wide range of species, including *Cajanus cajan, Lycopersicon esculentum, Nicotiana tabacum, Petunia hybrida, Triticum aestivum* and *Zea mays* (Bohjwani and Cocking, 1972; Bajaj, 1974; Deka, Mehra, Pathak and Sen, 1974). Tetrad protoplasts may be isolated from most fertile flowering plant species.

Within the *Nicotiana,* somatic hybrid plants have been recovered between nitrate reductase deficient (NR⁻) *N. tabacum* leaf meosphyll protoplasts (2n = 4x = 48) and *N. glutinosa* suspension cell protoplasts (2n = 2x = 24). The selection of hybrids was based on the restoration of nitrate reductase activity, green colour and regeneration capacity (Cooper-Bland *et al.,* 1985). To assess the use of tetrad protoplasts in fusion studies, this species combination was chosen.

RESULTS

Following fusion between nitrate reductase deficient (NR⁻) *N. tabacum* leaf mesophyll protoplasts, with *N. glutinosa* tetrad protoplasts, it was anticipated that only hybrid colonies would be capable of sustained division on a selection medium containing nitrate as sole nitrogen source. Two colonies were recovered in this way, one of which underwent regeneration giving rise to six shoots, five of which were successfully transferred to the greenhouse. The second colony was lost due to bacterial contamination.

The putative somatic hybrid plants displayed a number of morphological features characteristic of both fusion partners, and were found to have the expected amphipentaploid (but functionally amphitriploid) somatic chromosome complement of 3n = 5x = 60. Hybridity was confirmed on the basis of leaf esterase, and leaf callus peroxidase zymograms. The putative hybrids were found to possess a banding pattern which was the summation of the parental banding patterns for both iso-enzymes (see Fig. la and b), and included both *N. tabacum* and *N. glutinosa* specific bands.

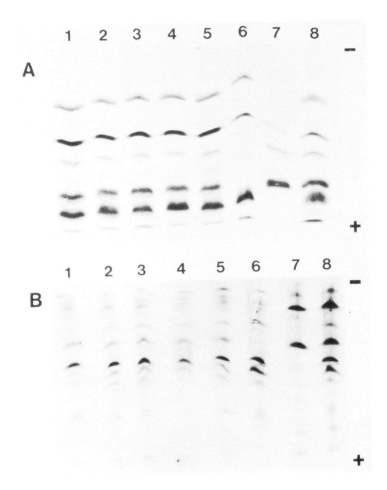

Fig. 1 - Isoelectric focussing of leaf callus peroxidases (A)
and leaf esterases (B) of the gametosomatic hybrids and
their parents. Channels from left to right: 1-5, triploid
gametosomatic hybrids; 6, *Nicotiana tabacum*; 7, *N. glutinosa*
and 8, physical mix (1:1) of proteins from *N. tabacum*
and *N. glutinosa*.

Analysis of the polypeptide composition of Fraction I protein extracted
from leaves of the five hybrids, and from *N. tabacum* and *N. glutinosa*
revealed that the hybrids possess the large subunit (chloroplast encoded)
of *N. tabacum*, and small subunits (nuclear encoded) of both *N. tabacum* and
N. glutinosa (see Fig. 2).

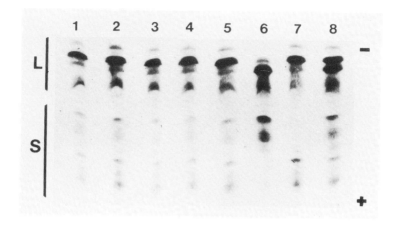

Fig. 2 - Isoelectric focussing of Fraction I protein
polypeptides from the gametosomatic hybrids and their
parents. Channels from left to right: 1-5, triploid
gametosomatic hybrids; 6, *Nicotiana glutinosa*; 7, N.
tabacum and 8, physical mix of proteins (1:1) from
N. tabacum and *N. glutinosa*. Hybrids have the large
subunit (L) of *N. tabacum* and small subunits (S) of both
N. tabacum and *N. glutinosa*.

Tentoxin sensitivity tests were performed with germinating seeds recovered following crosses between the five hybrids (female) and *N. tabacum*
(male) using the methods of Durbin and Uchytil (1977). The five somatic
hybrids gave rise to offspring which were resistant to tentoxin, characteristic of the *N. tabacum* chloroplast type. *N. glutinosa* was found to be
tentoxin sensitive.

DISCUSSION

This is the first demonstration that tetrad protoplasts can be used
successfully in fusion studies. Since somatic hybridisation describes
fusion between protoplasts derived from somatic cells, and tetrads are not
strictly somatic cells, we have proposed that novel triploid hybrids recovered following fusion between somatic cell protoplasts, and tetrad
protoplasts should be termed 'gametosomatic' hybrids to indicate this
distinction.

302

Tetrad protoplasts possess a number of characteristics which make them particularly suitable for fusion studies. Tetrads are an alternative source of protoplasts isolated directly from the plant, and as such are not subject to any effect due to long term tissue culture, which may be the case for protoplasts of cell suspension or callus origin. Tetrad protoplasts have been reported to be highly fusogenic (Bhojwani and Cocking, 1972). It has been proposed that a greater throughput of somatic hybrids may be achieved by using highly fusogenic protoplasts, since higher fusion frequencies may be obtained using less damaging fusion treatments (Boss, Grimes and Brightman, 1984). Since tetrad protoplasts have not been reported to undergo sustained division, there is no requirement to develop selection schemes against the tetrad protoplast, simplifying the selection of hybrid colonies or plants.

Somatic hybrids have previously been recovered between *N. tabacum* and *N. glutinosa* and were found to possess either the *N. glutinosa* chloroplast type (Uchimaya, 1982; Horn, Kameya, Brotherton and Widholm, 1984; Cooper-Bland, Pental and Cocking, 1985) or the *N. tabacum* chloroplast type (Cooper-Bland *et al.*, 1985). The results obtained from Fraction I protein analysis, and tentoxin sensitivity tests, indicate that the five gametosomatic hybrids reported here only possess functional *N. tabacum* chloroplasts. Many factors are thought to influence the sorting out of chloroplast types following protoplast fusion, including the physiological state of the donor protoplasts, the relative contribution from each protoplast source, as well as any inherent biological incompatibility (Fluhr, 1983). Somatic hybrids between *N. tabacum* and *N. glutinosa* have been recovered which possess either chloroplast type, indicating that strong biological incompatibilities do not exclude particular nuclear-cytoplasmic combinations (Cooper-Bland *et al.*, 1985). Cytoplasmic changes occurring during meiosis result in the dedifferentiation of both mitochondria and chloroplasts in the pollen mother cells (Heslop-Harrison, 1977). It might have been predicted that this would favour the retention of the *N. glutinosa* chloroplast type in the gametosomatic hybrids. However, this was not the case, although it must be stressed that data is only available at present for one hybrid event. Further gametosomatic hybrid colonies have been recovered, and a more detailed analysis of the chloroplast types found in a population of gametosomatic hybrids should clarify the extent to which the use of tetrad protplasts influences the subsequent sorting out of chloroplast types following protoplast fusion.

CONCLUSIONS

Tetrad protoplasts can be used in protoplast fusion studies, and novel gametosomatic hybrids recovered. The physiological status of the tetrad protoplast may differ quite considerably from protoplasts derived from other sources, and this may influence the nuclear/cytoplasmic recombination and/or segregation which occurs following protoplast fusion.
The limited data available suggest that the dedifferentiated plastids of the tetrad protoplasts do not have a selective advantage over the plastid contribution from a leaf mesophyll protoplast. This conclusion must be verified by the analysis of a larger population of gametosomatic hybrid plants and such a study is already underway.

ACKNOWLEDGEMENTS

The author wish to thank Dr. D. Pental for valuable discussions. We are grateful to Mr. I. Gilder, Mrs. J. Raynor and Mr. D Wilson for technical assistance, and Mr. B.V. Case for photography.

REFERENCES

Bajaj, Y.P.S. (1974). Isolation and culture studies on pollen mother cell-protoplasts. *Plant Sci. Lett. 3,* 93-99.

Bennett, M.D. (1976). In *Cell division in higher plants,* (ed. Yeoman, M.M. (1976). Academic Press, London 161-199.

Bhojwani, S.S., and Cocking, E.C. (1972). Isolation of protoplasts from pollen tetrads. *Nature, New Biol. 239,* 29-30.

Boss, W.F., Grimes, H.D., and Brightman, A. (1984). Calcium-induced fusion of fusogenic wild carrot protoplasts. *Protoplasma 120,* 209-215.

Cooper-Bland, S., Pental, D., and Cocking, E.C. (1985). Somatic hybridisation between nitrate reductase deficient *N. tabacum* and *N. glutinosa*. (In preparation).

Deka, P.C., Mehra, A.K., Pathak, N.N., and Sen, S.K. (1977). Isolation and fusion studies on protoplasts from pollen tetrads. *Experimenta 33,* 182-184.

Durbin, R.D., and Uchytil, J.F. (1977). A survey of plant insensitivity to tentoxin. *Phytopathology, 67,* 602-603.

Fluhr, R., (1983). The segregation of organelles and cytoplasmic traits in higher plant somatic fusion hybrids. *Experienta Supplementum, 46,* 85-92.

Heslop-Harrison, J., (1977). In *Pollen: Development and Physiology* (ed. Heslop-Harrison, J. (1977). Butterworths, London. 16-21.

Horn, M.E., Kameya, T., Brotherton, J.E., and Widholm, J.M., (1983). The use of aminoaacid analogue resistance and plant regeneration ability to select somatic hybrids between *Nicotiana tabacum* and *N. glutinosa.* *Mol. Gen. Genet. 192,* 235-240.

Maheshwari, S.C., Rashid,A., and Tyagi, A.K., (1982). Haploids from pollen grains - retrospect or prospect. *Amer. J. Bot. 69,* 865-879.

Negrutui, I., (1980). Nutritional requirements of protoplast derived haploid cells grown at low densities in liquid medium. *Planta 149,* 7-18.

Pental, D., and Cocking, E.C., (1985). Some theoretical and practical possibilities of plant genetic manipulation using protoplasts. *Hereditas Suppl. 3,* 83-93.

Uchimiya, H., (1982). Somatic hybridisation between male sterile *Nicotiana tabacum* and *N. glutinosa* through protoplast fusion. *Theor. Appl. Genet. 61,* 69-72.

Index

Index